OBJECTIVE QUESTIONS (MCQ) IN

HUMAN BIOCHEMISTRY

OBJECTIVE QUESTIONS (MCQ) IN HUMAN BIOCHEMISTRY

[For BSc, MSc, BPhram, BDS, MBBS and MD Level of all Indian Universities]

Dr Srikrushna Mahapatra, MD
Professor and Head, Department of Biochemistry,
MKCG Medical College, Berhampur, Orissa

New Central Book Agency (P) Ltd
LONDON
HYDERABAD ERNAKULAM BHUBANESWAR
DELHI KOLKATA PUNE GUWAHATI

NCBA

REGD OFFICE
8/1 Chintamoni Das Lane, Kolkata 700 009, India
ncbapvtltd@eth.net

OVERSEAS
NCBA (UK) Ltd, 149 Park Avenue North, Northampton, NN3 2HY, UK
ncbauk@yahoo.com

EXPORT
NCBA Exports Pvt. Ltd, 212 Shahpur Jat, New Delhi 110 049
ncbaexports@ncbaindia.com

BRANCHES

208B Peacock Lane, Shahpur Jat
New Delhi 110 049
ncbadel@ncbapvtltd.com

Shop Nos. 3 & 4, Vinayak Towers
681/B Budhwar Peth, Appa Balwant Chowk
Pune 411 002
ncbapune@ncbapvtltd.com

Shop No. 15, Bharati Towers, Ground Floor
Forest Park, **Bhubaneswar** 751 001
ncbabub@ncbapvtltd.com

House No. 3-1-315, 1st Floor
Nimboliadda, Kachiguda, **Hyderabad** 500 027
ncbahydb@ncbapvtltd.com

GSS Shopping Complex, 1st Floor
Opposite College Post Office, Convent Road
Ernakulam 682 035
ncbaernk@ncbapvtltd.com

Radhamayee Complex (Second Floor)
Opp. SBI, Panbazar, Jaswanta Road
Guwahati 781 001
ncbaguwahati@ncbapvtltd.com

Objective Questions (MCQ) in Human Biochemistry • S Mahapatra

First Published: 2006
Reprinted: January 2008, 2013

PUBLISHER AND TYPESETTER
New Central Book Agency (P) Ltd
8/1 Chintamoni Das Lane, Kolkata 700 009

PRINTER
New Central Book Agency (P) Ltd
Web-Offset Division, Dhulagarh, Sankrail, Howrah

COVER DESIGNER
Soumen Paul

TECHNICAL EDITOR
Dr Mita Sen

PROJECT TEAM
Partha Biswas and Prabhat Jas

ISBN 978 81 7381 497 6

Price ₹ 150.00

Dedicated to My Parents

CONTENTS

CONTENTS

PREFACE

Twentieth century saw exponential growth in the field of Biochemistry. The study of the blue print of human genome, a finding termed as *'Holy Grail of Human Genetics'* by the turn of the century opened the floodgate of knowledge and research in molecular biology. Knowledge in biology stands incomplete without in-depth study of biochemistry. Keeping pace with ever increasing demand on expertised personnel in biochemistry, the subject is taught in most of the universities of India at **B.Sc., M.Sc., B.Pharm, B.D.S., M.B.B.S., and MD level**. The present book aims at fulfilling demands of students preparing to undertake different entrance examinations. Care has been taken to cover all major topics in biochemistry on a broad base. The effort of the author would be worthwhile if it satisfies the requirement of students in their endeavour while preparing for different examinations.

I am thankful to my wife and daughter who have supported me during preparation of this manuscript. I acknowledge the help of Manoj Kumar Mohanty for the typing of the manuscript.

Suggestion for improvement and critical assessment from different quarters shall always be appreciated.

January, 2006
Berhampur, Orissa

Srikrushna Mahapatra

01 BIOCHEMICAL METHODS AND WATER

Q. 1. Which one of the following elements constitutes maximum elementary composition of human body on dry weight basis?

(a) Carbon **Answer** : (a)
(b) Oxygen
(c) Hydrogen
(d) Nitrogen.

Q. 2. Which one of the following chemical compounds contributes maxi-mum to the chemical composition of a normal man weighing 65 kgs?

(a) Protein **Answer** : (a)
(b) Fat
(c) Carbohydrate
(d) Mine.

Q. 3. Which one is the commonly used solution for extraction of sub-cellular organelle?

(a) SMTK **Answer** : (c)
(b) SKMT
(c) STKM
(d) MSTK.

Q. 4. Which of these is used as a market of lysosomes during subcellular organelle separation process?

(a) Alkaline phosphatase **Answer** : (d)
(b) Peroxidase
(c) Catalase
(d) Acid phosphatase.

Q. 5. During differential centrifugation of subcellular fractions which one separate out at low speed centrifugation?

(a) Microsomal fraction **Answer** : (c)
(b) Mitochondrial fraction
(c) Nuclear fraction
(d) Cytosolic fraction.

Q. 6. During differential centrifugation of subcellular fraction which statement is true.

(a) Very high speed centrifugation is required to separate mitocondrial fraction **Answer** : (c)
(b) Very high speed centrifugation is required to separate nuclear fraction
(c) Mitochondria fraction contain mitochondria, lysosome and peroxisome
(d) Microsomal fraction first separate out at low speed centrifugation.

Q. 7. The most detailed information about the structure of biomolecules is provided by

(a) Ion exchange chromatography **Answer** : (b)
(b) X-ray diffraction & Crystallography
(c) High pressure liquid chromatography
(d) SDS-PAGE.

Q. 8. Which of the statement is not true.

(a) Water molecules form dipoles **Answer** : (d)
(b) Hydrophobic interactions means self association of nonpolar compounds in aqueous medium
(c) Electrostatic interactions of protein molecules may be termed salt bridges
(d) Electrostatic interactions are strengthened by water.

Q. 9. Which of the bond exhibit highest bond energy?

(a) O-O **Answer** : (c)
(b) S-S
(c) N-H
(d) C-O.

Q. 10. Which type of bond exhibits lowest bond energy?

(a) O-O **Answer** : (a)
(b) C-H
(c) C-N
(d) O-H.

Q. 11. Which of the statements is true?

(a) Nonpolar hydrocarbon groups lack the capacity to form hydrogen bonds **Answer** : (a)
(b) Apolar molecules cannot form spherical droplets
(c) In water, hydrocarbons cannot form rigid clathrate structure
(d) Presence of apolar molecules decrease entropy.

Q. 12. Which one is a monoprotic acid?

(a) Lactic acid **Answer** : (a)
(b) Phosphoric acid
(c) Carbonic acid
(d) Succinic acid.

Q. 13. Which of these is a diprotic acid?

(a) Acetic acid **Answer** : (c)
(b) Lactic acid
(c) Carbonic acid
(d) Citric acid.

02 PROTEIN, HEMOGLOBIN AND ENZYMES

Q. 14. The natural amino acids which occur in protein are usually

(a) D-α Amino acid **Answer :** (b)

(b) L-α Amino acid

(c) D-β Amino acid

(d) L-β Amino acid.

Q. 15. Which one has an hydroxyl group in its side chain?

(a) Glycine **Answer :** (c)

(b) Alanine

(c) Serine

(d) Leucine.

Q. 16. Which is the most basic amino acid?

(a) Glycine **Answer :** (c)

(b) Alanine

(c) Arginine

(d) Lysine.

Q. 17. Amino acids are soluble in which one of the following solvents?

(a) Benzene **Answer :** (c)

(b) Hexane

(c) Ethanol

(d) Ether.

Q. 18. Which one of the amino acids is hydrophobic?

(a) Arginine **Answer :** (c)

(b) Glycine

(c) Alanine

(d) Glutamine.

Q. 19. Which of these is a hydrophilic amino acid?

(a) Methionine **Answer :** (d)

(b) Phenylalanine

(c) Alanine

(d) Lysine.

Q. 20. Which is untrue for amino acids?

(a) Amino acids have high melting point **Answer :** (b)

(b) Amino acids absorb visible light

(c) Histidine has imidazole nucleus

(d) Proline is an imino acid.

Q. 21. Which is most appropriate for a tripeptide?

(a) It has three amino acids only **Answer :** (d)
(b) It has three peptide bonds only
(c) It has three amino acids & three peptide bonds
(d) It has three amino acids & two peptide bonds.

Q. 22. Which is true for Glutathione?

(a) Glutathione is made up of four amino acids **Answer :** (c)
(b) Glutathione has glycine, methionine and glutamic acid
(c) Glutathione has a non-α-peptide bond
(d) Glutathione has two-α-peptide bonds.

Q. 23. Which statement is true?

(a) Peptide are not poly-electrolytes **Answer :** (b)
(b) Glutamate cyclised to pyroglutamic acid is TRH
(c) Glutathione has no role in metabolism of Xenobiotics
(d) Primary structure of protein cannot be determined by Edman's technique.

Q. 24. Who was first to sequence a polypeptide?

(a) F. Sanger **Answer :** (a)
(b) Edman
(c) Watson & Crick
(d) Nirenberg.

Q. 25. Which is not true for sequencing of amino acids?

(a) Sanger used 1-fluoro—2, 4 dinitrobenzine **Answer :** (d)
(b) In Sanger's method N—terminal amino acids residues are identified one by one
(c) Edman used phenyl iso thiocyanate to sequence amino acids
(d) Edman approach removes C-terminal amino acids.

Q. 26. In addition to Edman's technique, which other method can sequence peptides upto 25 amino acids in length?

(a) HPLC **Answer :** (b)
(b) Fast atom bombardment with mass spectrometry
(c) NMR and FAB
(d) Pulse field gel electrophoresis with Chromatography.

Q. 27. Who revolutionised the technique of protein synthesis?

(a) Emil Fisher **Answer :** (d)
(b) M. Brown
(c) F. Sanger
(d) B. Merrifield.

Q. 28. Which one of the following diseases is characterised by alteration in secondary—tertiary structure of protein?

(a) Scrapie **Answer :** (a)
(b) Pellagra
(c) Hurler's syndrome
(d) Thyrotoxicosis.

Q. 29. Which of these statements is wrong?

(a) Albumin is a globular protein **Answer : (c)**
(b) Histones are basic proteins
(c) Globular proteins have axial ratio more than 10
(d) Myoglobin is a storage protein.

Q. 30. Pick out the correct match

(a) Transport protein—Ferritin **Answer : (d)**
(b) Protective protein—Myoglobin
(c) Structural protein—Actin
(d) Motile protein—Tubulin.

Q. 31. Which one is a true statement, in relation to amino acid/protein.

(a) Angle between α-carbon & nitrogen is called Phi angle (ϕ) **Answer : (a)**
(b) Angle between α-carbon & carbonyl carbon is called Phi angle (ϕ)
(c) Angle between carbonyl carbon & nitrogen is called Phi (ϕ) angle
(d) Phi or Psi angle have nothing to do with Protein structure.

Q. 32. Which is untrue for an α-helix protein structure?

(a) There are 3·6 residues per turn **Answer : (b)**
(b) Amins acyl R groups are directed inward towards the helical axis
(c) On an average there are 12 amino acids per turn of an helix
(d) Pitch of the helix is about 5·4 A°.

Q. 33. Select out the incorrect statement.

(a) Hydrogen bonds stabilise α-helical structure of protein **Answer : (b)**
(b) Peptide nitrogen acts as hydrogen acceptors
(c) α-helix can form spontaneously
(d) Presence of glycine produces bend in protein structure.

Q. 34. Typical protein crystals have

(a) 10—20% water (b) 20—30% water **Answer : (c)**
(c) 40—60% water (d) 70—80% water.

Q. 35. Which of these techniques provide structure of protein in solution?

(a) X-ray crystallography **Answer : (b)**
(b) NMR spectroscopy
(c) Column chromatography
(d) High voltage electrophoresis.

Q. 36. In quaternary structure of protein, the association of subunits are stabilised by

(a) Covalent bonds **Answer : (d)**
(b) Disulphide bonds
(c) Peptide bonds
(d) Hydrogen bonds & electrostatic bonds.

Q. 37. Which of these set help in protein folding process?

(a) Protein disulfide isomerase, proly cis-trans isomerase, chaperonins **Answer : (a)**
(b) Protein monosulphide isomerase, prolyl cis-trans isomerase, chaperonins
(c) Chaperonins, protein kinases, protein disulfide isomerases
(d) Heat shock protein, cytokinins & protein kinases.

Q. 38. Which statement regarding prion protein is true?

(a) PrP-c is Protease insensitive **Answer** : (c)
(b) PrP-sc is Protease sensitive
(c) PrP-c & PrP-sc have identical primary structure
(d) PrP-c & PrP-sc have identical quaternary structure.

Q. 39. Which amino acid constitutes nearly 85% of silk fibroin?

(a) Proline **Answer** : (b)
(b) Glycine
(c) Hydroxyproline
(d) Alanine.

Q. 40. The primary structural motif of mature collagen is

(a) $(Gly—X—Pro/Hyp)_n$ **Answer** : (a)
(b) $(Pro—X—Hyp/Gly)_n$
(c) $(Gly—Pro—Hyp/Ala)_n$
(d) $(Hyp—Pro—X/Gly)_n$.

Q. 41. Which vitamin helps in hydroxylation of proline & lysine during collagen synthesis?

(a) Vit E **Answer** : (b)
(b) Vit C
(c) Vit D
(d) Vit B_{12}.

Q. 42. Which of these statements is untrue?

(a) Ehlers—Danlos syndrome is associated with defects in collagen maturation **Answer** : (d)
(b) Scurvy is associated with defects of collagen maturation
(c) X-ray crystallography study is useful to understand secondary structure of protein
(d) Urea cannot break non-covalent bonds of protein.

Q. 43. Affinity of Carbon monoxide to bind to isolated heme is how many times more than that of oxygen?

(a) 1,000 times **Answer** : (d)
(b) 5,000 times
(c) 20,000 times
(d) 25,000 times.

Q. 44. At what pO_2, myoglobin of muscle can release its bound oxygen for oxidative synthesis of ATP?

(a) 100 mm of Hg **Answer** : (d)
(b) 50 mm of Hg
(c) 10 mm of Hg
(d) 5 mm of Hg.

Q. 45. Pick out the incorrect match

(a) HbA-$\alpha_2\beta_2$ **Answer** : (c)
(b) HbS-$\alpha_2 S_2$
(c) HbA_2-$\alpha_2\gamma_2$
(d) α-Thalassemia—reduction in synthesis of α-chain.

Q. 46. Which of the statement is true for Hemoglobin?

(a) Hemoglobin is not a tetrameric protein **Answer :** (d)
(b) α-Peptide of hemoglobin has 146 amino acids
(c) β-Peptide of hemoglobin has 141 amino acids
(d) β-Peptide of HbA & myoglobin exhibit almost identical secondary & tertiary structure.

Q. 47. As regards oxygenation of hemoglobin which statement is untrue?

(a) Binding of oxygen accompanies rupture of salt bonds between carboxyl terminal residues in all four sub units **Answer :** (c)
(b) Conformational change occurs in hemoglobin molecule on oxygenation
(c) Deoxygenated hemoglobin is called R (relaxed) state
(d) These is profound alternation in secondary, tertiary & quaternary structure of hemoglobin on oxygenation.

Q. 48. Which of the statements is untrue?

(a) About 15% of the CO_2 carried in blood is carried directly on hemoglobin molecule **Answer :** (d)
(b) Amino terminal group of Hb reacts with CO_2 to form carbonate
(c) Hemoglobin binds two protons for every four oxygen molecules released
(d) Like hemoglobin, myoglobin also exhibits Bohr effect.

Q. 49. Pick out the incorrect statement.

(a) An increase in proton concentration promotes oxygen release **Answer :** (c)
(b) An increase in PO_2 promotes proton release
(c) The P_{50} of HbA is less than that of HbF
(d) Bohr effect is dependent on heme-heme interaction.

Q. 50. Which statement is untrue for 2-3 BPG?

(a) One molecule of 2-3 BPG is bound per hemoglobin tetramer **Answer :** (c)
(b) BPG is bound in the central cavity formed by all four subunits
(c) BPG stabilises the oxygenated state of hemoglobin
(d) BPG binds more weakly to foetal hemoglobin.

Q. 51. Select out the incorrect statement with regard to adoption to high altitude.

(a) There is an increase in number of erythrocytes **Answer :** (c)
(b) There is an increase in concentration of hemoglobin
(c) There is a decrease in concentration of 2-3 BPG
(d) There is an increase in concentration of 2-3 BPG.

Q. 52. Pick out the incorrect statement.

(a) In HbM, Tyr replaces His **Answer :** (c)
(b) In HbS, Valine replaces Glutamate at 6th position in β-chain
(c) In HbS, Glutamate replaces valine at 6th position in β chain
(d) In hemoglobin chesapeake R form exhibit increased oxygen affinity.

Q. 53. Which amino acid of hemoglobin typically helps in binding of heme iron with oxygen?

(a) Histidine (c) Valine **Answer :** (a)
(b) Leucine (d) Methionine.

Q. 54. What percent of total carbon dioxide of venous blood is carried by hemoglobin in the form of hemoglobin carbamate?

(a) 5% **Answer** : (c)
(b) 10%
(c) 15%
(d) 20%.

Q. 55. Find the incorrect statement.

(a) Coenzymes may be considered as second substrate **Answer** : (c)
(b) Digestive enzymes may be considered as prosthetic groups
(c) Digestive enzymes do require coenzymes
(d) Coenzymes are organic compounds.

Q. 56. Which coenzyme does not transfer hydrogen?

(a) NAD^+ **Answer** : (d)
(b) FAD
(c) Coenzyme Q
(d) Cobamide coenzymes.

Q. 57. Which enzyme does not exhibit absolute optical specificity?

(a) Oxidoreductases **Answer** : (b)
(b) Epimerases
(c) Hydrolases
(d) Ligases.

Q. 58. Which of the statement is in correct?

(a) Oxidation of NADH to NAD^+ is accompanied by rise of optical density at 340 mm **Answer** : (a)
(b) Reduction of NAD^+ to NADH is associated with rise in optical density at 340 mm
(c) Oxidation of NADH to NAD^+ is accompained by decrease of optical density at 340 mm
(d) One unit of enzyme activity refers to one micro mole substrate reacting or product transformed per minute.

Q. 59. In normal condition, which LDH isoenzyme is predominantly found in serum?

(a) LDH—1 **Answer** : (b)
(b) LDH—2
(c) LDH—3
(d) LDH—4.

Q. 60. In liver disease which LDH isoenzyme is abnormally raised?

(a) LDH—5 **Answer** : (a)
(b) LDH—4
(c) LDH—3
(d) LDH—1.

Q. 61. Which is a functional plasma enzyme?

(a) Lipoprotein lipase **Answer** : (a)
(b) Alkaline phosphatase
(c) Amylase
(d) Lipase.

Q. 62. Hexokinase belongs to which class of enzymes?

(a) Oxidoreductases **Answer : (b)**
(b) Transferases
(c) Ligase
(d) Lyase.

Q. 63. Which of the enzyme belongs to aspartic protease family of enzymes?

(a) Pepsin **Answer : (a)**
(b) Elastase
(c) Chymotrypsin
(d) Trypsin.

Q. 64. Which group belongs to serine protease family of enzyme?

(a) Pepsin, chymotrypsin, elastase **Answer : (c)**
(b) Pepsin, trypsin, elastase
(c) Trypsin, chymotrypsin, elastase
(d) Elastase, amylase, lipase.

Q. 65. Which of these statements is untrue for ΔG° (free energy under standard condition)?

(a) ΔG° = RT In Keq, where R is gas constant and 'I' is absolute temperature **Answer : (a)**
(b) ΔG° is independent of mechanism by which reaction proceeds
(c) ΔG° gives no information about rates of reaction
(d) ΔG° is negative if products have lower free energy than reactant.

Q. 66. Which statement is correct for enzyme catalysed reaction?

(a) Enzyme-substrate transition states are at lower energy level **Answer : (a)**
(b) Presence of enzyme alters ΔG° for overall reaction
(c) Enzymes can influence direction of a chemical reaction
(d) Enzymes can alter equilibrium constant of a chemical reaction.

Q. 67. Who proposed the 'Lock & Key' analogy of enzyme substrate inter-action?

(a) Koshland **Answer : (b)**
(b) Emil Fischer
(c) S. Ochoa
(d) A. Kornbeng.

Q. 68. Who proposed induced fit model of enzyme catalysis?

(a) Koshland **Answer : (a)**
(b) Emil Fischer
(c) Racker
(d) Messelsen.

Q. 69. Which one is Michaelis Menton expression?

(a) $V_i = \frac{V_{max}[S]}{K_m + [S]}$ (b) $K_m = \frac{V_{max}}{V_i[S]}$ **Answer : (a)**

(c) $V_{max} = \frac{V_i[S]}{K_m + [S]}$ (d) $K_m = \frac{V_i[S]}{V_{max}}$.

Q. 70. Graphic representation of Hill equation is helpful

(a) To evaluate sigmoid saturation kinetics **Answer : (a)**
(b) To evaluate Km/Vmax slope by straight line
(c) To evaluate half maximal velocity using Michaelis constant
(d) To evaluate Km value only.

Q. 71. A competitive inhibitor

(a) Decreases the Km of the substrate **Answer : (b)**
(b) Raises apparent Km for the substrate
(c) Has no effect on Km value
(d) Decreases apparent Km for the substrate.

Q. 72. Which is not a feature of reversible noncompetitive inhibitor?

(a) Inhibitor has little or no structural similarity with substrate **Answer : (c)**
(b) It lowers Vmax & usually do not affect Km
(c) If affects always the Vmax & Km
(d) Formation of both Enz SI & Enz SI possible.

Q. 73. Out of all the enzymes known, what % of enzymes have tightly bound metal ions, or require them for activity?

(a) 15% **Answer : (a)**
(b) 25%
(c) 35%
(d) 45%.

Q. 74. Which complex cannot be formed by metallo enzymes?

(a) Enz—S—M **Answer : (a)**
(b) M—Enz—S
(c) Enz—M—S
(d) $Enz<\begin{matrix}M\\ |\\ S\end{matrix}$

Q. 75. Which is untrue as regards the function of metal ions in enzymes?

(a) Accelerate general acid-base catalysis **Answer : (b)**
(b) Accelerate covalent catalysis
(c) Approximation of reactions
(d) Non-induction of strain on substrate.

Q. 76. Pick out the wrong match.

(a) Yersinia pestis—Protein tyrosine phosphatase **Answer : (b)**
(b) Vibio cholerae—Adenylation of M protein
(c) Oncogenes—encode Protein tyrosine kinase
(d) Lovastatin—3 HMG CoA reductase.

Q. 77. Who proposed the concept of homeostatic regulation of internal milieu?

(a) Bernard **Answer : (a)**
(b) Bergelius
(c) Henderson
(d) Sorensen.

Q. 78. Which one of the statements is incorrect?

(a) True equilibrium where bidirectional flow take place is not characteristic of life **Answer : (c)**
(b) Isomerases & Mutases often work unidirectionally in the cell due to siphoning of product
(c) Hydrolysis of protein to amino acids is not thermo-dynamically spontaneous
(d) Unidirectional nature of metabolic pathways renders them functionally specific.

Q. 79. Which of these enzymes is usually not listed as an inducible enzyme in eukaryotes?

(a) Threonine dehydratase **Answer : (d)**
(b) HMG CoA reductase
(c) Tryptophan pyrrolase
(d) Hexokinase.

Q. 80. Who deduced that body proteins are in a state of dynamic equilibrium, using ^{15}N—labelled amino acids?

(a) Schoenheimer **Answer : (a)**
(b) S. Ochoa
(c) Lehninger
(d) Sanger.

Q. 81. The kinetics of feed back inhibition may be the following types except.

(a) Competitive **Answer : (d)**
(b) Uncoupled
(c) Non-competitive
(d) Slightly irreversible.

Q. 82. Which is untrue for allosterism?

(a) Allosteric sites can be selectively denatured by urea. & X-rays **Answer : (b)**
(b) Allosteric effectors do not protect the catalytic site under any circumstance
(c) Structure of allosteric site & catalytic site are genetically distinct
(d) Allosteric & substrate may bind independently of each other.

Q. 83. Which is untrue for allosterism?

(a) In K-series allosteric enzymes, allosteric effector does not affect Vmax **Answer : (c)**
(b) In V-series allosteric enzymes, allosteric effector does not affect Km
(c) In K-series, binding of allosteric effector reduces the charge of catalytic residues
(d) In K-series, binding of effector may also weaken bonds between substrate & substrate binding residues.

Q. 84. Which one of the type of covalent modification is most widely used to regulate protein function?

Answer : (d)
(a) Glycosylation
(b) Adenylation
(c) Methylation
(d) Phosphorylation—dephosphorylation.

Q. 85. Which of these enzymes exhibit low activity on phosphorylation?

(a) Acetyl CoA carboxylase **Answer : (a)**
(b) Citrate lyase
(c) HMG CoA reductase kinase
(d) Phosphorylase kinase.

Q. 86. Which of these enzymes exhibit high activity on phosphorylation?

(a) Glycogen synthase **Answer : (d)**
(b) Pyruvate dehydrogenase
(c) HMG CoA reductase
(d) Citrate lyase.

Q. 87. Which statement is not correct?

(a) Change in free energy (ΔG) is also known as chemical potential in chemical systems **Answer : (d)**
(b) If ΔG is negative, reaction is exergonic
(c) If ΔG is positive, reaction is endergonic
(d) Entropy is minimum in a system as it approaches true equilibrium.

Q. 88. Who introduced the concept of high energy phosphate?

(a) Lipmann **Answer : (a)**
(b) Lehninger
(c) Lohman
(d) F. Racker.

Q. 89. Which one has the most standard free energy of hydrolysis?

(a) Phosphoenolpyruvate **Answer : (a)**
(b) Creatine phosphate
(c) ATP
(d) Pyrophosphate.

Q. 90. Which one has the least standard free energy of hydrolysis?

(a) ADP **Answer : (d)**
(b) Glucose-1-phosphate
(c) Fructose-6-phosphate
(d) AMP.

Q. 91. Which one of the enzyme listed below represents 'X' in the reaction ATP + AMP $\xleftrightarrow{\text{'X'}}$ 2ADP?

(a) Myokinase **Answer : (a)**
(b) Phosphate kinase
(c) Adenosine kinase
(d) Nucleoside diphosphate kinase.

Q. 92. Which one of the following has least standard free energy of hydrolysis?

(a) Carbamoyl phosphate **Answer : (d)**
(b) Phosphoenolpyruvate
(c) Creatine phosphate
(d) ATP.

Q. 93. Which one of these can act as a phosphagen?

Answer : (c)

(a) Glucose-6-phosphate
(b) ADP
(c) Creatine phosphate
(d) Glycerol phosphate.

Q. 94. Which one of these enzymes is FMN linked?

Answer : (c)

(a) Glucose oxidase
(b) Aldehyde dehydrogenase
(c) L—Amino acid oxidase
(d) Succinate dehydrogenase.

Q. 95. Which is not true for catalase?

Answer : (d)

(a) It uses hydrogen peroxide as an electron donor and electron acceptor
(b) It has peroxidase activity
(c) It contains four heme groups
(d) Under most conditions its peroxidase activity is not favoured.

03 BIOENERGETICS AND METABOLISM OF CARBOHYDRATES

Q. 96. Which of these enzymes incorporate oxygen into the substrate?

(a) Oxidases **Answer** : (b)
(b) Oxygenases
(c) Dehydrogenases
(d) Catalases.

Q. 97. Which is not correct for mitochondrial cytochrome P_{450} mon-oxygenase system?

(a) They help in hydroxylation reaction during steroidogenesis **Answer** : (c)
(b) In adrenal cortex, mitochondrial cytochrome P_{450} is six times more abundant then cytochromes of respiratory chain
(c) In liver it helps I-α hydroxylation of 25-hydroxycholecalciferol
(d) It helps in 26-hydroxylation of bile acid synthesis.

Q. 98. Which one of these is not a dioxygenase enzyme?

(a) Homogentisate oxidase **Answer** : (d)
(b) Tryptophan pyrrolase
(c) 3-hydroxy anthranilate oxidase
(d) L—Amino acid oxidase.

Q. 99. Which one of the enzymes is not a marker of outer mitochondrial membrane?

(a) Monoamine oxidase **Answer** : (d)
(b) Phospholipase A_2
(c) Acyl CoA synthetase
(d) Adenylyl kinase.

Q. 100. Pick out the incorrect match

(a) Succinate dehydrogenase—inner surface of inner mitocondrial membrane **Answer** : (b)
(b) 3—hydroxy butyrate dehydrogenase—outer membrane of mitochondria
(c) Glycerol-3-phosphate dehydrogenase—outerside of inner mitochondrial membrane
(d) Creatine kinase—inter membranal space of mitochondria.

Q. 101. Which structure of mitochondria resembles vit-K?

(a) Coenzyme Q **Answer** : (a)
(b) Cytochrome C
(c) Cytochrome B
(d) F_0-F_1 complex.

Q. 102. Select the correct sequence of components in respiratory chain from among the choices given.

(a) NAD → Fp.FeS → Q → Cytb → $Cytc_1$ → $Cytaa_3$ → O_2 **Answer : (a)**

(b) NAD → FAD → Q → Cyt, b → CytC, C → Cyt. aa_3 → O_2

(c) NAD → Q → Fp.FeS → Cyt C, C → Cytb → $Cylaa_3$ → O_2

(d) NAD → Fp.FeS → FAD → Q → Cytc, C → Cytb → $Cytaa_3$ → O_2.

Q. 103. Which group is more mobile component of inner mitochondrial membrane among the choices given?

(a) Cytochrome C & Q **Answer : (a)**

(b) Cytochrome b & Q

(c) Cytochrome b & Cytochrome C

(d) Cytochrome a – a_3.

Q. 104. Which one of the enzymes has FAD as the coenzyme?

(a) Malate dehydrogenase **Answer : (d)**

(b) 3 hydroxy acyl CoA dehydrogenase

(c) Isocitrate dehydrogenase

(d) Succinate dehydrogenase.

Q. 105. Cells in resting condition are in which state of respiratory control as detailed by Chance & Williams?

(a) Stage 1 **Answer : (d)**

(b) Stage 2

(c) Stage 3

(d) Stage 4.

Q. 106. Pickout the incorrect match as regards states of respiratory control as described by Chance & William.

(a) Stage 1—Availability of ADP & substrate **Answer : (d)**

(b) Stage 2—Availability of substrate only

(c) Stage 3—Capacity of respiratory chain itself

(d) Stage 4—Availability of oxygen only.

Q. 107. Which one inhibits complex IV of respiratory chain in mitochondria?

(a) BAL **Answer : (d)**

(b) Rotenone

(c) Carboxin

(d) Hydrogen sulphide.

Q. 108. Who proposed the chemiosmotic theory of oxidative phosphorylation?

(a) Mitchell **Answer : (a)**

(b) F. Racker

(c) Lehninger

(d) S. Ochoa.

Q. 109. Which is a competitive inhibitor of succinate dehydrogenase?

(a) Malonate **Answer : (a)**

(b) Malate

(c) Dinitrophenol

(d) Barbiturate.

Q. 110. Which is not an uncoupler of oxidative phosphorylation?

(a) 2-4 Dinitrophenol **Answer :** (b)
(b) Atractyloside
(c) Dinitrocresol
(d) CCCP.

Q. 111. Which of the statement is incorrect?

(a) Valinomycin allows penetration of potassium ions through mitochondrial membrane and acts as an ionophore **Answer :** (d)
(b) Nigericin also act as an ionophore
(c) Dinitrophenol is a protein ionophore
(d) Ionophores alter only the membrane potential and not the pH gradient across the membrane.

Q. 112. Inherited defect of complex IV or cytochrome oxidase of respiratory chain produces?

(a) Weber's disease **Answer :** (b)
(b) MELAS
(c) Patau syndrome
(d) Splenic aplasia.

Q. 113. Orientation of –H and –OH groups about which carbon of glucose makes it either D or L form?

(a) 1st carbon atom **Answer :** (c)
(b) 3rd carbon atom
(c) 5th carbon atom
(d) 6th carbon atom.

Q. 114. Crystalline glucose is

(a) α-D—Glucopyranose **Answer :** (a)
(b) β-D—Glucopyranose
(c) α-L—Glucopyranose
(d) β-L—Glucopyranose.

Q. 115. Glucose in solution, approximately has a mixture of

(a) 38% α-Glucopyranose and 62% β-Glucopyranose **Answer :** (a)
(b) 62% α-Glucopyranose and 38% β-Glucopyranose
(c) 50% each of α and β Glucopyranose
(d) 25% β-Glucopyranose and 75% α-Glucopyranose.

Q. 116. Galatose is an epimer of glucose around which carbon of glucose?

(a) 1st Carbon **Answer :** (d)
(b) 2nd Carbon
(c) 3rd Carbon
(d) 4th Carbon.

Q. 117. Which one of these is found in urine in essential pentosuria?

(a) Ribose **Answer :** (d)
(b) Ribulose
(c) Xylose
(d) Xylulose.

Q. 118. Which one of these is a fructosan?

(a) Starch **Answer :** (c)
(b) Cellulose
(c) Inulin
(d) Chitin.

Q. 119. β-L-Fucose is a

(a) Deoxy galactose **Answer :** (a)
(b) Deoxy glucose
(c) Deoxy fructose
(d) Deoxy mannose.

Q. 120. Neuraminic acid is a nine carbon sugar derived from

(a) Galactosamine and Pyruvate **Answer :** (b)
(b) Mannosamine and Pyruvate
(c) Fructosamine and Pyruvate
(d) Fucosamine and Pyruvate.

Q. 121. Counted from the carboxyl end and inclusive of the carboxyl carbon of a fatty acid, β-caron is?

(a) 1st Carbon **Answer :** (c)
(b) 2nd Carbon
(c) 3rd Carbon
(d) 4th Carbon.

Q. 122. Which one is a 18 carbon, mono unsaturated fatty acid?

(a) Palmitoleic acid **Answer :** (b)
(b) Oleic acid
(c) Stearic acid
(d) Erucic acid.

Q. 123. Which one of the fatty acids has six double bonds?

(a) Cervonic acid **Answer :** (a)
(b) Clupanodonic acid
(c) Timnodonic acid
(d) Arachidonic acid.

Q. 124. Cephalin contains which one of these compounds?

(a) Serine **Answer :** (b)
(b) Ethanolamine
(c) Lecithin
(d) Inositol.

Q. 125. Ceramide is made up of?

(a) Sphingosine and fatty acid **Answer :** (a)
(b) Sphingosine and amino acid
(c) Cetyl alcohol and cholesterol
(d) Sphingosine and choline.

Q. 126. Which glycolipid is predominantly found in brain and nervous tissue?

(a) Galactosyl ceramide **Answer :** (a)
(b) Glucosyl ceramide
(c) Mannosyl ceramide
(d) Fructosyl ceramide.

Q. 127. Cholesterol is not a precursor of which one these compounds?

(a) Bile acids **Answer** : (d)
(b) Vitamin D
(c) Cardiac glycosides
(d) Adrenocorticotrophic hormone.

Q. 128. Which one is a hexaenoic acid?

(a) Elaidic acid **Answer** : (d)
(b) Arachidonic acid
(c) Timnodonic acid
(d) Cervonic acid.

Q. 129. Which is a major lipid of mitochondrial membrane?

(a) Cardiolipin **Answer** : (a)
(b) Ceramide
(c) Sphingomyeline
(d) Plasmalogen.

Q. 130. Which is not a correct statement?

(a) Carbohydrate, fat and protein after digestion & their respective metabolic processes produce a common product acetyl CoA **Answer** : (b)
(b) End product of digestion of carbohydrate, protein & fat form assimilatory substances such as glucose, amino acids & chylomicron which directly go to liver
(c) Free fatty acids are not taken up by brain
(d) Surplus triacylglycerol from lipogenesis is secreted as VLDL into circulation.

Q. 131. Which is not true for mitochondria?

(a) It has enzymes of citric acid cycle **Answer** : (b)
(b) It is the only site of ATP generation
(c) It is the site of β-oxidation of fatty acid
(d) It is site of ketone body synthesis.

Q. 132. Which contains the enzymes of acylglycerol synthesis?

(a) Endoplasmic reticulum **Answer** : (a)
(b) Mitochondria
(c) Avilable free in cytosol
(d) Golgi complex.

Q. 133. Which is not a correct statement?

(a) In metabolic process, non equilibrium reaction are potential control points **Answer** : (d)
(b) Flux-generating reaction is usually a reaction saturated with substrate
(c) Allosteric control is an important control mechanism in enzyme catalysed reaction
(d) Hormones do not have much importance in enzyme-catalysed reaction.

Q. 134. CoA is produced from which vitamin?

(a) Pyridoxine **Answer** : (b)
(b) Pantothenic acid
(c) Biotin
(d) Lipoic acid.

Q. 135. Which statement is untrue for TCA cycle?

(a) Anoxia or hypoxia causes inhibition of TCA cycle **Answer :** (c)
(b) Two molecules of CO_2 is released per cycle
(c) Usually equal or large amounts of oxaloacetate is needed as compared to acetyl CoA
(d) Oxaloacetate plays a catalytic role.

Q. 136. Among the following reaction, which one is considered physiologically unidirectional?

(a) Citrate to Cis—Aconitate **Answer :** (d)
(b) Cis—Aconitate to Isocitrate
(c) Isocitrate to α-Ketoglutarate
(d) α-Ketoglutarate to succinyl CoA.

Q. 137. Which is untrue for succinate dehydrogenase enzyme?

(a) It is present in the matrix of mitochondria **Answer :** (a)
(b) It transfers reducing equivalents from substrate to flavoproteins directly
(c) Addition of either malonate or oxaloacetate may inhibit succinate dehydrogenase
(d) Succinate is converted to fumarate by this enzyme.

Q. 138. How many soluble vitamin of B complex have precise role in citric acid cycle?

(a) Two (c) Four **Answer :** (c)
(b) Three (d) Five.

Q. 139. Isoleucine and valine may contribute to gluconeogenesis, as their carbon skeleton can enter TCA cycle at the step of?

(a) Succinyl CoA **Answer :** (a)
(b) α-ketoglutarate
(c) Malate
(d) Oxaloacetate.

Q. 140. Serine and glycine may contribute to gluconeogenesis as their carbon skeleton can form?

(a) Formate **Answer :** (d)
(b) Oxaloacetate
(c) Malate
(d) Pyruvate.

Q. 141. Which of these pathways is amphibolic in nature?

(a) Glycolysis **Answer :** (b)
(b) TCA cycle
(c) Pentose phosphate pathway
(d) Glycogenesis.

Q. 142. Which is not true for glycolysis?

(a) It is the major pathway of glucose utilisation **Answer :** (b)
(b) It is not the main pathway of fructose and galactose metabolism
(c) It is found in cytosol of all cells
(d) It can operate in the presence or absence of oxygen.

Q. 143. Hexokinase can be allosterically inhibited by?

(a) Glucose 6 phosphate **Answer** : (a)
(b) Glucose 1 phosphate
(c) ATP
(d) ADP.

Q. 144. Which is not true for hexokinase enzyme?

(a) It has low Km for glucose **Answer** : (b)
(b) It cannot act on α and β anomers of glucose
(c) It can catalyse phosphorylation of other hexoses
(d) It is allosterically inhibited by Glu 6 phosphate.

Q. 145. If lactate is the end product of glycolysis, the net ATP produced is?

(a) 1 **Answer** : (b)
(b) 2
(c) 4
(d) 8.

Q. 146. Pick out the set of enzymes which are the major sites of regulation of glycolysis.

(a) Hexokinase, phosphofructokinase, pyruvate kinase **Answer** : (a)
(b) Phosphofructokinase, phospho glycerate kinase, glyceraldehyde 3 phosphate dehydrogenase
(c) Phosphofructokinase, phosphoglycerate kinase, pyruvate kinase
(d) Phosphofructokinase, glyceraldehyde 3 phosphate dehydrogenase, pyruvate kinase.

Q. 147. Which is not true for pyruvate dehydrogenase complex enzyme?

(a) If require 3 coenzymes **Answer** : (c)
(b) It converts pyruvate to acetyl CoA
(c) NADH stimulates the activity of pyruvate dehydrogenase complex
(d) Acetyl CoA inhibits pyruvate dehydrogenase complex.

Q. 148. How many high energy phosphates are formed per mole of glucose under aerobic conditions, when metabolised in humans?

(a) 30 **Answer** : (b)
(b) 38
(c) 34
(d) 40.

Q. 149. Typically deficiency of which enzyme of glycolytic pathway produces hemolytic anemia?

(a) Pyruvate kinase **Answer** : (a)
(b) Enolase
(c) Phosphofructokinase
(d) Phosphoglycerate kinase.

Q. 150. The commonly used chemical sodium fluoride inhibits which enzyme of glycolysis?

(a) Hexokinase **Answer** : (c)
(b) Pyruvate kinase
(c) Enolase
(d) Phosphofructokinase.

Q. 151. During glycogen synthesis which molecule contributes C-1 for the 1-4 glycosidic bond formation?

(a) UDPGIc **Answer :** (a)
(b) Glu-1-phosphate
(c) Terminal glucose residue of glycogen
(d) Glu-6-phosphate.

Q. 152. Glycogenin is a

(a) Protein **Answer :** (a)
(b) Oligomeric carbohydrate
(c) Glycolipid
(d) Lipoprotein.

Q. 153. Which is the rate limiting step of glycogenolysis?

(a) Step catalysed by glucan transferase **Answer :** (b)
(b) Step catalysed by phosphorylase
(c) Step by amylo 1→6 glucosidase
(d) Branching enzyme.

Q. 154. Which is not true?

(a) Epinephrine stimulates adenyl cyclase acting through α-adrenergic receptor **Answer :** (a)
(b) Epinephrine stimulate adenyl cyclase acting through β-adrenergic receptor
(c) Glucagon stimulates adenyl cyclase
(d) Epinephrine and Norepinephrine both stimulate adenyl cyclase.

Q. 155. Which one is caused by deficiency of glucose-6-phosphatase enzyme?

(a) Von Gierke's disease **Answer :** (a)
(b) Pompe's disease
(c) Cori's disease
(d) Her's disease.

Q. 156. Which one is present in muscle phosphorylase?

(a) Lipoic acid **Answer :** (b)
(b) Pyridoxal phosphate
(c) Tetrahydrofolate
(d) Pantothenate.

Q. 157. Activity of which enzyme is reduced in starvation?

(a) Pyruvate kinase **Answer :** (a)
(b) Pyruvate carboxylase
(c) Glucose-6-phosphatase
(d) Fructose-1-6 biphosphatase.

Q. 158. Which is not a substrate for Gluconeogenesis?

(a) Alanine **Answer :** (c)
(b) Lactate
(c) Acetyl CoA
(d) Propionate.

Q. 159. Fructose 1-6 biphosphatase enzyme is believed to be absent in which tissue?

(a) Liver **Answer :** (d)
(b) Kidney
(c) Striated muscle
(d) Smooth muscle.

Q. 160. Deficiency of which vitamin may lead to methyl malonic aciduria?

(a) Vit B_6 **Answer :** (b)
(b) Vit B_{12}
(c) Folic acid
(d) Niacin.

Q. 161. Insulin may act as an inducer of which enzyme?

(a) Pyruvate carboxylase **Answer :** (d)
(b) Hexokinase
(c) Fructose 1-6 bisphosphatase
(d) Glucokinase.

Q. 162. Which type of glucose transporter system is insulin dependent?

(a) Glut-1 **Answer :** (d)
(b) Glut-2
(c) Glut-3
(d) Glut-4.

Q. 163. Increased tolerance to glucose is seen in

(a) Diabetes mellitus **Answer :** (c)
(b) Thyrotoxicosis
(c) Adrenocortical insufficiency
(d) Gigantism.

Q. 164. The enzyme which converts propionyl CoA to D-methyl malonyl CoA requires coenzymic form of which vitamin?

(a) Thiamin **Answer :** (c)
(b) Niacin
(c) Biotin
(d) Folic acid.

Q. 165. Methyl malonyl CoA isomerase requires coenzyme of which vitamin?

(a) Pyridoxal phosphate **Answer :** (c)
(b) Folic acid
(c) B_{12}
(d) Pantothenate.

Q. 166. Which one of these is used in parenteral nutrition, but in high concentration can cause hepatic necrosis?

(a) Fructose **Answer :** (a)
(b) Galactose
(c) Glucose
(d) Mannose.

Q. 167. Which statement is true for pentose phosphate pathway?

(a) From one mole of Glu. 6. phosphate, 6 moles of NADPH is generated **Answer : (c)**
(b) From three moles of Glu. 6. phosphate, 3 moles of glyceraldehyde-3-phosphate generated
(c) From three moles of Glu. 6 phosphate, 3 moles of CO_2 is generated
(d) From one mole of Glu. 6 phosphate 3 moles of CO_2 is generated.

Q. 168. In pentose phosphate pathway by the enzymatic action of 3-epimerase, ribulose-5-phosphate is converted to

(a) Ribose-5-phosphate **Answer : (b)**
(b) Xylulose-5-phosphate
(c) Xylose-5-phosphate
(d) Sedoheptulose-5-phosphate.

Q. 169. Transketolase enzyme requires which coenzyme?

(a) Pyridoxal phosphate **Answer : (b)**
(b) Thiamine diphosphate
(c) Thiamine triphosphate
(d) Pantothenic acid.

Q. 170. Which is not true for pentose phosphate pathway?

(a) It is not active in nonlactating mammary gland **Answer : (c)**
(b) It is active in thyroid gland
(c) It is highly active in skeletal muscles
(d) It is highly active in adipose tissues.

Q. 171. Which of these statements is not true?

(a) It is necessary to have a completely functioning pentose phosphate pathway for a tissue to synthesize ribose phosphate **Answer : (a)**
(b) Skeletal tissue has very small amounts of glucose-6-P-dehydrogenase
(c) Pentose phosphate pathway provides NADPH
(d) Pentose phosphate pathway is active in RBC.

Q. 172. Which of the enzymes have selenium?

(a) Glutathion reductase **Answer : (b)**
(b) Glutathion peroxidase
(c) Glue-6-phosphate dehydrogenase
(d) Epimerase.

Q. 173. Absence of which one of these enzymes makes humans incapable of synthesizing ascorbic acid?

(a) L—Gulonolactone oxidase **Answer : (a)**
(b) L—Gluconolactone oxidase
(c) Ascorbate oxidase
(d) L—Gulonolactone reductase.

Q. 174. Which is true?

(a) 3-Keto—L-Gulonate is oxidised to L-Gulonate **Answer : (d)**
(b) Guinea pigs can synthesize vitamin C
(c) UDP—Glucuronate is not required in proteoglycan synthesis
(d) Fructose is more rapidly glycolysed by liver than glucose.

Q. 175. Fructokinase converts fructose to which one metabolite with the help of ATP?

(a) Fructose—6—phosphate **Answer** : (b)
(b) Fructose—1—phosphate
(c) Fructose—2—phosphate
(d) Fructose—5—phosphate.

Q. 176. Which of these statements is true?

(a) Glucose inhibits phosphorylation of fructose when both are present in the medium **Answer** : (a)
(b) Galactose is essential in diet
(c) Hexokinase does not phosphorylate fructose
(d) Aldolase B does not split fructose—1—phosphate.

Q. 177. Which sugar predominantly appears in urine in essential pentosuria?

(a) L—Xylulose **Answer** : (a)
(b) L—Xylose
(c) D—Ribose
(d) L—Ribose.

Q. 178. Hereditary fructose intolerance is due to absence of which enzyme?

(a) Fructokinase **Answer** : (b)
(b) Aldolase B
(c) Aldolase C
(d) Hexokinase.

04 METABOLISM OF LIPIDS

Q. 179. Where does chain elongation of fatty acids takes place?

(a) Mitochondria
(b) Golgi complex
(c) Lysosome
(d) Endoplasmic reticulum.

Answer : (d)

Q. 180. Which one of these is not required for de novo synthesis of fatty acids?

(a) NADPH
(b) Biotin
(c) HCO_3^- as a source of CO_2
(d) NADH.

Answer : (d)

Q. 181. Which vitamin is part of acyl carrier protein (ACP)?

(a) Pantothenic acid
(b) Biotin
(c) Thiamin
(d) Pyridoxal phosphate.

Answer : (a)

Q. 182. Which one is not a component of fatty acid synthase complex?

(a) Acetyl CoA carboxylase
(b) Ketoacyl synthase
(c) Acetyl transacylase
(d) Hydratase.

Answer : (a)

Q. 183. Which is untrue for lipogenesis?

(a) Dietary fat depresses lipogenesis
(b) Lipogenesis is higher when glucose is fed rather than sucrose
(c) Nutritional state of an organism regulate rate of lipogenesis
(d) Deficiency of insulin depress lipogenesis.

Answer : (b)

Q. 184. Which one of these doesnot suppress lipogenesis?

(a) Restricted caloric intake
(b) High fat diet
(c) Deficiency of insulin
(d) Sucrose in diet.

Answer : (d)

Q. 185. Detect the incorrect statement.

(a) Acetyl CoA carboxylase is an allosteric enzyme
(b) Insulin stimulates acetyl CoA carboxylase
(c) cAMP stimulates acetyl CoA carboxylase
(d) Long chain acyl CoA inhibit acetyl CoA carboxylase enzyme.

Answer : (c)

Q. 186. Which is not true for acyl CoA synthetase?

(a) Acyl CoA synthetases are found on outer mitochondrial membrane
(b) They are found in peroxisomes
(c) They are not found in endoplasmic reticulum
(d) Specific acyl CoA synthetases are there for fatty acids of different chain lengths.

Answer : (c)

Q. 187. From which two amino acids carnitine is synthesized?

(a) Glycine & methionine
(b) Glutamate & methionine
(c) Valine & methionine
(d) Lysine & methionine.

Answer : (d)

Q. 188. Which is the coenzyme for the enzyme L(+)—3—hydroxy acyl CoA dehydrogenase?

(a) NAD^+
(b) FAD
(c) NADP
(d) FMN.

Answer : (a)

Q. 189. Which is not true for fatty acids?

(a) Propionyl CoA is produced from β-oxidation of odd chain fatty acids
(b) Propionyl CoA is glucogenic
(c) Short, medium & long chain fatty acids, all require carnitine transporter system
(d) Free fatty acid is bound to albumin in plasma.

Answer : (c)

Q. 190. Which is not true for fatty acid oxidation?

(a) End product of long chain fatty acid oxidation in peroxisomes is acetyl CoA & H_2O_2
(b) Enzymes of peroxisome attack short chain fatty acids to oxidise them quickly
(c) Peroxisomes do not contain carnitine palmitoyl transferase
(d) Side chain cleavage of cholesterol can occur in peroxisomes.

Answer : (b)

Q. 191. The normal ketone bodies level in blood in well fed humans is

(a) 0·2 mmol/L
(b) 0·6 mmol/L
(c) 1 mmol/L
(d) 2 mmol/L.

Answer : (a)

Q. 192. Which is the predominant ketone body in blood & urine in ketosis?

(a) Aceto acetic acid
(b) 3 hydroxy butyric acid
(c) Acetone
(d) Acetic acid.

Answer : (b)

Q. 193. How many ATPs are produced from 1 mole of palmitate if 3-hydroxy butyrate is the end product via β-oxidation?

(a) 129
(b) 33
(c) 29
(d) 21.

Answer : (d)

Q. 194. Jamaican vomiting sickness produced due to eating of unripe fruit of akee true is due to

(a) Hypoglycin
(b) Glycirrhyza
(c) Toxolysin
(d) Hydrolysine.

Answer : (a)

Q. 195. Deficiency of medium chain acetyl CoA dehydrogenase produces

(a) Dicarboxylic aciduria
(b) Monocarboxylic aciduria
(c) Zellweger's syndrome
(d) Refsum's disease.

Answer : (a)

Q. 196. How many carbons are present in eicosanoic acids?

(a) 18
(b) 20
(c) 22
(d) 24.

Answer : (b)

Q. 197. At which position, usually the first double bond is introduced into a saturated fatty acid in our system?

(a) Δ^4
(b) Δ^5
(c) Δ^6
(d) Δ^9.

Answer : (d)

Q. 198. Which one is an omega-3 fatty acid?

(a) Linoleic acid
(b) α-linolenic acid
(c) γ-linolenic acid
(d) Arachidonic acid.

Answer : (b)

Q. 199. Which one corresponds to palmitoleate?

(a) 16 : 1, ω-7
(b) 16 : 1, ω-6
(c) 18 : 1, ω-7
(d) 18 : 1, ω-6

Answer : (a)

Q. 200. Which one is untrue for desaturation and chain elongation process?

(a) Diminished in fed state
(b) Diminished in fasting state
(c) Diminished upon glucagon administration
(d) Diminished in absence of insulin.

Answer : (a)

Q. 201. Decosahexaenoic acid (DHA), a very important fatty acid for development of retina and brain, can be synthesized from

(a) α-Linolenic acid
(b) γ-Linolenic acid
(c) Arachidonic acid
(d) Linoleic acid.

Answer : (a)

Q. 202. Which of the statements is incorrect?

(a) Prostaglandins, leukotriens, thromboxanes & lipoxins, all are eicosanoids
(b) Indomethacin inhibit cycloxygenases by competing with arachidonate
(c) Prostaglandins cannot relieve symptoms of essential fatty acid deficiency
(d) Essential fatty acid deficiency syndrome can be caused by chronic inhibition of prostaglandin synthesis.

Answer : (d)

Q. 203. Which leukotriene is first formed in lipoxygenase pathway?

(a) Leukotriene A_4
(b) Leukotriene B_4
(c) Leukotriene C_4
(d) LXA_4.

Answer : (a)

Q. 204. Which compound is necessary to produce leukotriene C_4 from leukotriene A_4?

(a) Methionine
(b) Glutathione
(c) Oxygen
(d) Reduced NAD^+.

Answer : (b)

Q. 205. Platelet activating factor is an

(a) Nitro phospholipid
(b) Alkyl phospholipid
(c) Amino phospholipid
(d) Not a phospholipid.

Answer : (b)

Q. 206. Which phosphoglycerol is typically derived from dihydroxy acetone phosphate?

(a) Phosphatidyl choline
(b) Phosphatidyl ethanolamine
(c) Cardiolipin
(d) Plasmalogen.

Answer : (d)

Q. 207. Which phospholipid is specifically required in inner mitochondrial membrane for cytochrome oxidase activity?

(a) Phosphatidyl ethanolamine
(b) Lecithin
(c) Cardiolipin
(d) Phosphatidate.

Answer : (c)

Q. 208. Lysolecithin is produced from lecithin by the action of which enzyme?

(a) Phospholipase A_1
(b) Phospholipase A_2
(c) Phospholipase B
(d) Phospsholipase C.

Answer : (b)

Q. 209. Ceramide is made up of

(a) Amino acid and fatty acid **Answer** : (a)
(b) Amino acid only
(c) Fatty acid and glycogen
(d) Amino acid, fatty acid & glycerol.

Q. 210. Sphingomyelin synthesis occurs mainly in

(a) Endoplasmic reticulum **Answer** : (c)
(b) Cytosol
(c) Golgi apparatus
(d) Lysosome.

Q. 211. What is true of Lignoceric acid?

(a) Completely synthesized from acetyl CoA **Answer** : (a)
(b) Requires acetyl CoA and sugar
(c) Requires glycerol & acetyl CoA
(d) No synthesized from acetyl CoA.

Q. 212. Nervonic acid is

(a) Derived from lignoceric acid **Answer** : (d)
(b) Derived from cerebronic acid
(c) Poly unsaturated fatty acid derived from linolenic acid
(d) Derived from oleic acid.

Q. 213. ABO blood group substances are

(a) Simple phospholipids **Answer** : (c)
(b) Glyco phospholipids
(c) Glyco sphingolipids
(d) Phosphatidates.

Q. 214. Deficiency of which enzyme can produce Goucher's disease?

(a) Ceramide **Answer** : (c)
(b) Sphingomyelinase
(c) β-Glucosidase
(d) β-Galactosidase.

Q. 215. The X-linked recessive disorder, Fabry's disease is due to deficiency of which one of these?

(a) β-Glucosidase **Answer** : (b)
(b) α-Galactosidase
(c) Sphingomyelinase
(d) Ceramidase.

Q. 216. Which one of the following has least amount of protein?

(a) Chylomicron **Answer** : (a)
(b) Chylomicron remnant
(c) VLDL
(d) LDL.

Q. 217. Which of these HDL types has highest phospholipid content?

(a) HDL_1 (b) HDL_2 **Answer** : (d)
(c) HDL_3 (d) Pre β-HDL.

Q. 218. Which apolipoprotein acts as a cofactor for lipoprotein lipase?

(a) Apo A—I
(b) Apo A—II
(c) Apo C—I
(d) Apo C—II.

Answer : (d)

Q. 219. Which apolipoprotein acts as a cofactor for the enzyme LCAT?

(a) Apo A—I
(b) Apo A—II
(c) Apo C—I
(d) Apo B—II.

Answer : (a)

Q. 220. Which apolipoprotein acts as a ligand for LDL receptor?

(a) Apo A—I
(b) Apo B—100
(c) Apo C—I
(d) Apo—E.

Answer : (b)

Q. 221. Which apolipoprotein acts as a ligand for remnant receptor?

(a) Apo A—I
(b) Apo B—48
(c) Apo C—I
(d) Apo—E.

Answer : (d)

Q. 222. Which apoprotein acts as a ligand for HDL receptor?

(a) Apo B—48
(b) Apo B—100
(c) Apo A—I
(d) Apo—E.

Answer : (c)

Q. 223. Which statement is untrue for chylomicrons?

(a) Found in chyle of lymphatic system draining the intestine only
(b) Responsible for transport of all dietary lipids
(c) Formation does not occur in fasting state
(d) Chylomicron formation increases with the load of triacylglycerol absorbed.

Answer : (c)

Q. 224. Which carry most of the triacylglycerol from the liver to extrahepatic tissues?

(a) Chylomicron
(b) VLDL
(c) LDL
(d) HDL.

Answer : (b)

Q. 225. Which apolipoprotein is essential for chylomicron and VLDL synthesis?

(a) Apo A
(b) Apo B
(c) Apo C
(d) Apo E.

Answer : (b)

Q. 226. Lipoprotein lipase enzyme is inactive in

(a) Adipose tissue (b) Renal medulla **Answer** : (d)

(c) Neonatal liver (d) Adult liver.

Q. 227. Which statement is not true?

(a) Injection of heparin increases lipoprotein lipase activity in blood **Answer** : (c)

(b) Large quantities of heparin also increases hepatic lipase secretion

(c) Hepatic lipase readily reacts with chylomicron

(d) Hepatic lipase is concerned with HDL metabolism.

Q. 228. Which is not true for LDL?

(a) Most LDL appear to be formed from VLDL **Answer** : (d)

(b) LDL has apo B 100

(c) Carboxy terminal domain of Apo B 100 contain the ligand for LDL receptor

(d) Approximately 30% LDL is degraded in liver & 70% in extrahepatic tinue.

Q. 229. Which is not true for HDL?

(a) HDL acts as a repository of apo C and apo E **Answer** : (d)

(b) Nascent HDL consists of apo A and free cholesterol

(c) HDL_2 concentration is inversely related to the incidence of coronary artery disease

(d) Pre β-HDL is least potent in inducing cholesterol efflux from the tissues.

Q. 230. Which is not true about fatty livers?

(a) Carbon tetrachloride cause fatty liver **Answer** : (c)

(b) Metabolic block in production of plasma lipoprotein produce fatty liver

(c) Orotic acid prevents fatty liver by decreasing plasma lipoprotein containing apo B

(d) Choline helps in recovery from fatty liver.

Q. 231. Which is untrue for alcoholism?

(a) Ethanol consumption over a long period can cause fatty liver **Answer** : (c)

(b) Usually women have increased susceptibility to alcohol consumption

(c) Alcohol increases uric acid excretion

(d) Alcohol causes hyperlactic acidemia.

Q. 232. Which is not true for triacylglycerol metabolism?

(a) Triacylglycerol stores of adipose tissue undergo lipolysis and reesterification continuously **Answer** : (b)

(b) Glycerol kinase is active in adipose tissue, liver and kidney

(c) Hormone sensitive lipase is distinct from lipoprotein lipase

(d) Provision of glycerol—3-phosphate regulates esterification of fatty acids.

Q. 233. Which statement is incorrect?

(a) Insulin inhibits hormone sensitive lipase **Answer** : (d)
(b) Insulin enhances lipogenesis
(c) Insulin reduces free fatty acid level
(d) Pentose phosphate pathway is not affected by insulin action.

Q. 234. Which is not true for brown adipose tissue?

(a) Brown adipose tissue is reduced or absent in obese persons **Answer** : (b)
(b) It has high activity of ATP synthase
(c) It has high content of mitochondria
(d) It has well developed blood supply.

Q. 235. Cholesterol is synthesized from

(a) Acetyl CoA only **Answer** : (a)
(b) Acetyl CoA and propionyl CoA
(c) Acetyl CoA & Acyl CoA
(d) Acetyl CoA & malonyl CoA.

Q. 236. Which is not a correct statement?

(a) Cholesterol is an ampipathic lipid **Answer** : (c)
(b) Acetyl CoA is the source of all carbon atoms of cholesterol
(c) Liver accounts for 70% of total synthesis of cholesterol
(d) Intestine accounts for 10% of total synthesis of cholesterol.

Q. 237. Which intermediary product of cholesterol synthesis branches out to produce ubiquinone & dolichol?

(a) Mevalonate (b) HMG CoA **Answer** : (c)
(c) Farnesyl diphosphate (d) Geranyl diphosphate.

Q. 238. HMG CoA reductase is inhibited by

(a) Mevalonate (b) Acetyl CoA **Answer** : (a)
(c) Acetoacetyl CoA (d) HMG CoA.

Q. 239. Which statement is not true?

(a) Cholesterol inhibits HMG CoA reductase **Answer** : (d)
(b) Cholesterol synthesis exhibits diurnal variation
(c) Thyroid increases HMG CoA reductase activity
(d) Insulin decreases HMG CoA reductase activity.

Q. 240. Which lipoprotein has highest proportion of cholesterol?

(a) VLDL (b) IDL **Answer** : (c)
(c) LDL (d) HDL.

Q. 241. Usually how much of cholesterol is eliminated from the body of an healthy adult per day?

(a) 1 gm (b) 2 gm **Answer** : (a)
(c) 3 gm (d) 5 gm.

Q. 242. Which is not a correct statement?

(a) Bile acids are synthesized from cholesterol **Answer** : (b)
(b) 5 α-hydroxylation is the rate limiting step in biosynthesis of bile acids

(c) Vitamin C deficiency interferes with bile acid synthesis
(d) Bile acid synthesis requires NADPH & cytochrome P—450 system.

Q. 243. Where does conjugation of bile acids take place?

(a) Endoplasmic reticulum (b) Cytosol **Answer** : (d)
(c) Golgi complex (d) Peroxisomes.

Q. 244. Which one of these is least reabsorbed during enterohepatic circulation of bile acids?

(a) Cholic acid **Answer** : (d)
(b) Chenodeoxycholic acid
(c) Deoxycholic acid
(d) Lithocholic acid.

Q. 245. Which one of these increases LDL catabolism?

(a) Sitosterol (b) Lovastatin **Answer** : (d)
(c) Gemfibrozil (d) Probucol.

Q. 246. Deficiency of lysosomal cholesteryl ester hydrolase produces which one?

(a) Tangier's disease **Answer** : (c)
(b) Fish eye disease
(c) Wolman's disease
(d) Familial dysbetalipoproteinemias.

Q. 247. Which is not true of leptin?

(a) Leptin is a weight regulatory hormone **Answer** : (d)
(b) Stimulates lipolysis
(c) Inhibits lipogenesis
(d) Leptin is an enzyme which affects lipolysis & lipogenesis.

Q. 248. Which one of these stimulate HMG CoA reductase activity for cholesterol synthesis?

(a) Glucagon (b) Glucocorticoids **Answer** : (c)
(c) Insulin (d) Mevalonate.

Q. 249. Apo (a) shows structural homology to

(a) Plasminogen (b) Fibrinogen **Answer** : (a)
(c) Thromboxane (d) Prothrombin.

Q. 250. Which is a major regulatory enzyme of citric acid cycle?

(a) Pyruvate dehydrogenase **Answer** : (d)
(b) Aconitase
(c) α-Ketoglutarate dehydrogenase
(d) Citrate synthase.

Q. 251. Which is the major regulatory enzyme of glycolysis?

(a) Glu-6-phosphatase (b) Hexokinase **Answer** : (c)
(c) Phosphofructokinase (d) Pyruvate kinase.

Q. 252. Which is the major regulatory enzyme of lipogenesis?

(a) Ketoacyl synthase (b) Acetyl CoA carboxylase **Answer** : (b)
(c) ACAT (d) LCAT.

05 METABOLISM OF PROTEIN, AMINO ACIDS AND HEME

Q. 253. Which is not the usual major fuel of heart?

(a) Free fatty acid **Answer : (d)**
(b) Lactate
(c) Ketone bodies
(d) Amino acid.

Q. 254. Which may be considered specialised enzyme of heart?

(a) Lipoprotein lipase **Answer : (a)**
(b) Lactate dehydrogenase
(c) Glycerol kinase
(d) Glucokinase.

Q. 255. Which one is nutritionally essential amino acid from among the following?

(a) Glycine **Answer : (d)**
(b) Tyrosine
(c) Proline
(d) Lysine.

Q. 256. Which one is nutritionally non-essential amino acid from the choices given?

(a) Leucine **Answer : (d)**
(b) Methionine
(c) Phenylalanine
(d) Tyrosine.

Q. 257. Which statement is not true?

(a) Hydroxyl group of hydroxyproline is unsubstituted in collagen **Answer : (d)**
(b) Hydroxyl group of hydroxylysine are attached to glucosyl and galactosyl residues
(c) 5 hydroxylysine is absent in most mammalian protein
(d) Hydroxyl group of hydroxyproline is attached with glucose or galactose residues.

Q. 258. Which is true for selenocysteine?

(a) Selenocysteine is synthesized from cysteine **Answer : (c)**
(b) Codon for selenocysteine is AUA
(c) Selenocysteine has a specific UGA codon
(d) Selenocysteine cannot be incorporated cotranslationally.

Q. 259. Among the following enzymes which has shorter half life?

Answer : (d)

(a) Aldolase
(b) Lactate dehydrogenase
(c) Cytochromes
(d) Tyrosine transaminase.

Q. 260. Which is not true for protein degradation?

Answer : (c)

(a) Intracellular proteases hydrolyse internal peptide bonds
(b) Extracellular proteins are degraded in lysosomes
(c) Short lived proteins are degraded in cytosol and do not require ATP
(d) Protein degradation in lysosome is ATP independent.

Q. 261. Presence of which amino acid at amino terminal end accelerate protein degradation by ubiquitin?

Answer : (c)

(a) Methionine
(b) Serine
(c) Aspartic acid
(d) No specific amino acid.

Q. 262. Which one of the following amino acid does not undergo transamination to remove α-amino nitrogen?

Answer : (d)

(a) Glutamate
(b) Arginine
(c) Histidine
(d) Lysine.

Q. 263. Which is not true for urea synthesis?

Answer : (c)

(a) Urea synthesis occurs in liver
(b) Synthesis of 1 mol of urea requires 3 mols of ATP
(c) Synthesis of 1 mol of urea requires 2 moles of ATP
(d) Synthesis of urea involves five enzymes.

Q. 264. Which acts as an enzyme activator in urea synthesis?

Answer : (a)

(a) N—acetyl glutamate
(b) N—acetyl aspartate
(c) Ornithine
(d) Citrulline.

Q. 265. Which one of the amino acid is consumed during urea synthesis?

Answer : (d)

(a) Arginine
(b) Citrulline
(c) Ornithine
(d) Aspartate.

Q. 266. Which is not true of ammonia production in kidney?

Answer : (a)

(a) Ammonia formation in tubular cells facilitate anion conservation
(b) Intra-cellular glutamine is broken down by glutaminase to release ammonia
(c) It helps in acid base regulation
(d) Ammonia production increases metabolic acidosis.

Q. 267. Which tissue generates more than half of the total body pool of free amino acids?

(a) Liver

(b) Muscle

(c) Intestine

(d) Skin.

Answer : (b)

Q. 268. In liver which serves as a key gluconeogenic amino acid?

(a) Alanine

(b) Glycine

(c) Serine

(d) Glutamic acid.

Answer : (a)

Q. 269. Which one can compete with arginine & inhibit arginase enzyme?

(a) Lysine

(b) Histidine

(c) Carbamoyl phosphate

(d) Citrulline.

Answer : (a)

Q. 270. From which amino acid Nitric Oxide (NO) can be synthesized?

(a) Proline (b) Valine

(c) Methionine (d) Arginine.

Answer : (d)

Q. 271. Which is the pacemaker enzyme of urea cycle?

(a) Arginase

(b) Ornithine transcarbamoylase

(c) Carbamoyl phosphate synthase—1

(d) Arginosuccinic acid synthase.

Answer : (c)

Q. 272. Which of these disorders is X-chromosome linked?

(a) Hyperammonemia—1

(b) Hyperammonemia—2

(c) Citrullinemia

(d) Argino succinicaciduria.

Answer : (b)

Q. 273. Which one is the ketogenic amino acid?

(a) Methionine (b) Proline

(c) Threonine (d) Leucine.

Answer : (d)

Q. 274. The carbon skeleton of asparate can enter the TCA cycle as which amphibolic intermediate?

(a) Fumarate

(b) Oxaloacetate

(c) Malate

(d) α-ketoglutarate.

Answer : (b)

Q. 275. Formiminoglutamate (FIGLU) is a metabolic product of which amino acid?

(a) Glycine

(b) Serine

(c) Histidine

(d) Arginine.

Answer : (c)

Q. 276. Which is not true for phenylketonuria (PKU)?

(a) It is due to deficiency of phenylalanine hydroxylase enzyme **Answer** : (c)
(b) Patients may give a mousy odor
(c) Diet containing low phenylalanine can be terminated at 3 years of age
(d) Screening of newborn infants for PKU is compulsory in United States.

Q. 277. From which amino acid nicotinic acid can be synthesized?

(a) Tryptophan **Answer** : (a)
(b) Tyrosine
(c) Histidine
(d) Valine.

Q. 278. Cystathionine is produced by

(a) Condensation of serine with homocysteine **Answer** : (a)
(b) Condensation of cysteine with serine
(c) Condensation of methionine with homocysteine
(d) Condensation of glycine with homocysteine.

Q. 279. Deficiency of which vitamin can produce methyl malonyl aciduria?

(a) Thiamine **Answer** : (d)
(b) Riboflavine
(c) Folic acid
(d) Vitamin B_{12}.

Q. 280. Sarcosine component of creatine has

(a) Glycine & S-adenosyl methionine **Answer** : (a)
(b) α-glutamate & glycine
(c) α-glutamate & S-adenosyl methionine
(d) Aspartate & glycine.

Q. 281. Which statement is not true?

(a) Carnosine is a β-alanyl dipeptide **Answer** : (d)
(b) β-alanine is present in coenzyme A
(c) β-alanine can arise from metabolite of pyrimidines in mammals
(d) Bacteria do not produce β-alanine.

Q. 282. Which statement is not correct?

(a) Anserine is methylcarnosine **Answer** : (b)
(b) Carnosine can activate myosin ATPase activity, but anserine does not
(c) Both carnosine & anserine chelate copper
(d) Homocarnosine is present in brain.

Q. 283. Which is not a histidine compound?

(a) Ergothionine **Answer** : (d)
(b) Anserine
(c) Carnosine
(d) Cystathionine.

Q. 284. Which amino acid forms polyamines such as spermine and spermidine?

(a) Ornithine (b) Lysine
(c) Leucine (d) Histidine.

Answer : (a)

Q. 285. From which amino acid melatonin is produced?

(a) Lysine (b) Leucine
(c) Arginine (d) Tryptophan.

Answer : (d)

Q. 286. Melanin is produced from which amino acid?

(a) Tyrosine
(b) Tryptophan
(c) Valine
(d) Leucine.

Answer : (a)

Q. 287. Which three amino acids are required for creatine synthesis?

(a) Glycine, cysteine, glutamate
(b) Arginine, cysteine, glycine
(c) Glycine, arginine, methionine
(d) Arginine, glutamate, methionine.

Answer : (c)

Q. 288. Metabolism of which one of these bases produces β-alanine?

(a) Thymine
(b) Cytosine
(c) Adenine
(d) Hypoxanthine.

Answer : (b)

Q. 289. Which amino acid acts as the precursor for synthesis of taurine?

(a) Glycine (b) Serine
(c) Threonine (d) Cysteine.

Answer : (d)

Q. 290. In addition to β-alanine which other amino acid is required for carnosine (dipeptide) synthesis?

(a) Lysine
(b) Arginine
(c) Histidine
(d) Cysteine.

Answer : (c)

Q. 291. Which one is anserine?

(a) N—methyl carnosine
(b) N—acetyl carnosine
(c) Acetylated spermine
(d) Deoxy spermine.

Answer : (a)

Q. 292. Which of these doesnot arise from ornithine?

(a) Spermine
(b) Spermidine
(c) Putrescine
(d) Anserine.

Answer : (d)

Q. 293. Which one of these is not a hemoprotein?

(a) Myoglobin
(b) Cytochrome
(c) Catalase
(d) Peroxidase.

Answer : (d)

Q. 294. Which amino acid acts as the starting material of heme synthesis?

(a) Glycine **Answer** : (a)
(b) Serine
(c) Threonine
(d) Glutamic acid.

Q. 295. ALA synthase-I, the rate limiting enzyme of heme synthesis requires which one that activates glycine in the reaction?

(a) Thiamine pyrophosphate **Answer** : (c)
(b) Riboflavin
(c) Pyridoxal phosphate
(d) Biotin.

Q. 296. Which is untrue for porphyrins?

(a) Porphyrin is a cyclic tetrapyrrole **Answer** : (b)
(b) Porphyrinogen & Porphyrins are both coloured compounds
(c) Porphyrinogens are reduced porphyrins
(d) Soret band is a sharp absorption band near 400 nm due to porphin ring.

Q. 297. Which enzymatic defect produces acute intermittent porphyria?

(a) ALA synthase **Answer** : (c)
(b) ALA dehydratase
(c) Uroporphyrinogen synthase-I
(d) Uroporphyrinogen decarboxylase.

Q. 298. Porphyrinogens do not flouresce due to

(a) Absence of double bonds joining pyrrole rings **Answer** : (a)
(b) Presence of methyl side chains
(c) Presence of vinyl side chains
(d) Presence of propionyl side chains.

Q. 299. Protoporphyria (erythropoeitic) is produced due to deficiency of which enzyme?

(a) ALA synthase **Answer** : (b)
(b) Ferro chelatase
(c) ALA dehydratase
(d) Coproporphyinogen oxidase.

Q. 300. One gm of hemoglobin usually yields how much bilirubin?

(a) 15 mg **Answer** : (b)
(b) 35 mg
(c) 55 mg
(d) 75 mg.

Q. 301. What is the daily bilirubin production in an healthy human adult?

(a) 100—150 mg% **Answer** : (b)
(b) 250—350 mg%
(c) 500—600 mg%
(d) 600—750 mg%.

Q. 302. Each molecule of albumin has how many binding sites for bilirubin?

(a) One high affinity site only

(b) One high affinity & one low affinity site

(c) Two high affinity sites

(d) Does not have any binding site.

Answer : (b)

Q. 303. Which is not true for bilirubin?

(a) Biliverdin is reduced to produce bilirubin

(b) In 100 ml of plasma, approximately 25 mg of bilirubin can be tightly bound to albumin

(c) Conjugated bilirubin is secreted into bile by a passive transport mechanism

(d) Bilirubin enter hepatocytes from plasma by a facilitated transport system.

Answer : (c)

Q. 304. Which of these statement is incorrect?

(a) Retention hyperbilirubinemia is usually unconjugated hyperbilirubinemia

(b) Unconjugated hyperbilirubinemia only produces kernicterus

(c) Regurgitation hyperbilirubinemia is usually conjugated hyperbilirubinemia

(d) Conjugated bilirubin being water soluble can enter blood brain barrier.

Answer : (d)

Q. 305. Which one produces conjugated hyperbilirubinemia?

(a) Neonatal physiologic jaundice

(b) Crigler—Najjar syndrome

(c) Gilbert syndrome

(d) Rotor syndrome.

Answer : (d)

06 PURINE, PYRIMIDINE AND NUCLEIC ACIDS

Q. 306. The site of attachment of ribose to pyrimidine bases is usually at which position?

(a) N—3 of Pyrimidine — **Answer : (b)**
(b) N—1 of Pyrimidine
(c) C—1 of Pyrimidine
(d) C—3 of Pyrimidine.

Q. 307. In pseudouridine, ribose is attached to uracil at

(a) N—1 of Uracil — **Answer : (d)**
(b) N—2 of Uracil
(c) C—1 of Uracil
(d) C—5 of Uracil.

Q. 308. Hypouricemia may be seen in which condition?

(a) Xanthine oxidase deficiency — **Answer : (a)**
(b) HGPRT deficiency
(c) Glu-6-phosphatase deficiency
(d) APRT deficiency.

Q. 309. Chemically adenine is what?

(a) 6—amino purine — **Answer : (a)**
(b) 2—amino oxy purine
(c) 4—amino pyrimidine
(d) 2—amino purine.

Q. 310. Among the choices given which one is a methylated base?

(a) Cytosine — **Answer : (c)**
(b) Uracil
(c) Thymine
(d) Guanine.

Q. 311. The most abundant free nucleotide in mammalian cell is

(a) ATP — **Answer : (a)**
(b) CTP
(c) UTP
(d) GTP.

Q. 312. Which acts as the second messenger for Nitric oxide (NO)?

(a) cAMP — **Answer : (b)**
(b) cGMP
(c) DAG
(d) Inosine triphosphate.

Q. 313. Which is the parent nucleotide from which both AMP and GMP are formed?

(a) Inosine monophosphate **Answer** : (a)
(b) Xanthosine monophosphate
(c) Orotidyl monophosphate
(d) Citidyl monophosphate.

Q. 314. Which three compounds are necessary for purine base synthesis?

(a) Glycine, lysine, histidine **Answer** : (b)
(b) Glycine, aspartate, glutamine
(c) Glycine, glutamine, histidine
(d) Glycine, histidine, proline.

Q. 315. Which is the major determinant of purine nucleotide biosynthesis?

(a) Phosphoribosyl pyrophosphate (PRPP) **Answer** : (a)
(b) Adenine phosphoribosyl transferase (APRT)
(c) Hypoxanthine—guanine phosphoribosyl transferase (HGPRT)
(d) Uric acid.

Q. 316. Which is not true for pyrimidine nucleotide synthesis?

(a) Pyrimidine base synthesis require aspartate, glutamine, CO_2 **Answer** : (d)
(b) Ribose phosphate moeity is attached later in the biosynthetic pathway
(c) UMP is the first true pyrimidine nucleotide synthesized
(d) UTP is aminated to CTP by aspargine.

Q. 317. Which is true for pyrimidine biosynthesis?

(a) The synthetic process is cytosolic only **Answer** : (c)
(b) There are five enzymatic steps
(c) Five enzymes of the synthetic pathway are organised as multifunctional polypeptide
(d) CMP is the first true pyrimidine nucleotide formed in the denovo biosynthetic process.

Q. 318. How much of uric acid is excreted in urine per day in normal human beings?

(a) 100—200 mg (b) 200—400 mg **Answer** : (c)
(c) 400—600 mg (d) 600—800 mg.

Q. 319. Which statement is not correct?

(a) Urates are more soluble in water than uric acid **Answer** : (d)
(b) Uricase cleaves uric acid to produce water soluble compound called allantion
(c) Uric acid stone formation can be reduced by alkalinisation of urine
(d) Sodium urate stone formation occurs distal to the site of urine acidification.

Q. 320. Lesch-Nyhan syndrome occurs due to definiency of

(a) PRPP synthetase **Answer** : (b)
(b) HGPRT
(c) Adenosine deaminase
(d) APRT.

Q. 321. Hypouricemia is produced in deficiency of .

(a) PRPP synthetase **Answer :** (c)
(b) Adenosine deaminase
(c) Xanthine oxidase
(d) Glu. 6 phosphatase.

Q. 322. Which deficiency of purine metabolism is X-linked recessive?

(a) HGPRT **Answer :** (a)
(b) Purine nucleoside phosphorylase
(c) ADA
(d) Xanthine oxidase.

Q. 323. Which inherited defect of pyrimidine metabolism is X-linked recessive?

(a) β-aminoisobutyric aciduria **Answer :** (d)
(b) Orotic aciduria type—I
(c) Orotic aciduria type—II
(d) Ornithine transcarbamoylase deficiency.

Q. 324. Which group of scientists demonstrated first that DNA contained genetic information?

(a) Avery, Macleod and McCarty **Answer :** (a)
(b) Watson and Crick
(c) Matthei, Nirenberg
(d) Nirenberg, H. Khorana.

Q. 325. How many hydrogen bonds are there between adenine and thymine?

(a) One **Answer :** (b)
(b) Two
(c) Three
(d) Four.

Q. 326. If there are 44 bases in total in a strand of DNA, & there are 11 adenine nucleotides, then how many thymine nucleotides are present?

(a) 11 (b) 22 **Answer :** (a)
(c) 33 (d) 12.

Q. 327. Which is untrue for DNA structure?

(a) Double stranded DNA can exist in six forms **Answer :** (d)
(b) Non-coding strand is the template strand
(c) DNA strands run antiparallel
(d) In low salt, high degree of hydration, 'A' form DNA is usually seen.

Q. 328. Which is truc for DNA?

(a) G—C bonds are more resistant to denaturation or melting **Answer :** (a)
(b) A—form of DNA is the usual physiological form
(c) Denaturation reduces optical absorbance of purine and pyrimidines
(d) The width of double helix in B form DNA is 4 mm.

Q. 329. Which RNA may have roles in RNA processing?

(a) hn RNA **Answer :** (d)
(b) m RNA
(c) t RNA
(d) Sn RNA.

Q. 330. Which class of RNA is the most heterogeneous in size & stability?

(a) m RNA **Answer :** (a)
(b) r RNA
(c) t RNA
(d) None of the above.

Q. 331. Which arm of tRNA is important for recognition of a tRNA by its specific aminoacyl tRNA synthetase enzyme?

(a) CCA arm **Answer :** (c)
(b) Anticodon arm
(c) D-arm
(d) Thymidine pseudouridine cytidine arm.

Q. 332. Which arm of tRNA helps in binding of amino acyl tRNA with ribosomal surface during protein synthesis?

(a) CCA arm **Answer :** (d)
(b) Anticodon arm
(c) D-arm
(d) Thymidine pseudouridine cytidine arm.

Q. 333. Which statement is true?

(a) Transcriptionally inactive chromatin stains less densely **Answer :** (c)
(b) Transcriptionally active chromatin stains more densely
(c) DNAse-I makes single strand cuts in any segment of DNA
(d) Heterochromatin is transcribed earlier to euchromatin in mammalian cell cycle.

Q. 334. Which statement is untrue/in correct?

(a) Constitutive heterochromatin is always condensed **Answer :** (d)
(b) Facultative heterochromatin may look like euchromatin
(c) One of the X-chromosomes out of the pair of X chromosome in mammalian family is hetrochromatic
(d) In early embryogenesis, heterochromatic X chromosome stays completely condensed.

Q. 335. Which statement is incorrect?

(a) Ends of chromosomes are called telomeric region **Answer :** (d)
(b) Centromere is an A-T rich region
(c) Centromere binds to several protein to form kinetochore complex
(d) Telomeric end consist of short repeat A-U rich sequences.

Q. 336. Human haploid genome consist of how many base pairs?

(a) 3×10^9 **Answer :** (a)
(b) 3×10^7
(c) 6×10^9
(d) 6×10^7.

Q. 337. Length of DNA is compressed about how many folds to produce a condensed metaphase chromosome?

(a) 5,000 **Answer** : (d)
(b) 6,000
(c) 7,000
(d) 8,000.

Q. 338. Fragile X syndrome is associated with which unstable repeat trinucleotide sequence?

(a) CGC **Answer** : (a)
(b) CAG
(c) CTG
(d) ATG.

Q. 339. Complete replication of lagging stand of DNA is carried out by.

(a) DNA polymerase I **Answer** : (a)
(b) DNA polymerase II
(c) DNA polymerase III
(d) DNA primase.

Q. 340. Which statement is incorrect?

(a) DNA replication is conservative **Answer** : (a)
(b) DNA is synthesized in $5' \rightarrow 3'$ direction
(c) The mobile complex between helicase and primase is called primosome
(d) Okazaki fragments are short newly synthesized DNA fragments in lagging strand.

Q. 341. If TGTAA is the sequence of bases in template strand in 5' to 3' direction during replication, which is the correct sequence in the new DNA strand?

(a) ACATT (b) TTACA **Answer** : (b)
(c) GCTAC (d) TTUCU.

Q. 342. Fault in which type of DNA repair has been linked to Hereditary Non-Polyposis Colon Cancer (HNPCC)?

(a) Mismatch repair **Answer** : (a)
(b) Base excision repair
(c) Nucleotide excision repair
(d) Double strand break repair.

Q. 343. Spontaneous deamination of adenine produces?

(a) Uracil **Answer** : (d)
(b) Guanine
(c) Xanthine
(d) Hypoxanthine.

Q. 344. Xeroderma pigmentosum, an autosomal recessive genetic disease is associated with fault in which repair process?

(a) Mismatch repair **Answer** : (c)
(b) Base excision repair
(c) Nucleotide excision repair
(d) Double strand break repair.

Q. 345. Which is not true for transcription (Process of RNA synthesis)?

(a) The steps of transcriptions are initiation, elongation, and termination **Answer** : (c)
(b) Ribonucleotides are required
(c) RNA polymerases require a primer
(d) There is no proof-reading in transcription.

Q. 346. Which is untrue for transcription?

(a) Only a small portion of genome is transcribed into RNA **Answer** : (c)
(b) The strand of DNA that is transcribed is called template strand
(c) Information in template strand is read in 5 to 3 direction
(d) RNA polymerases do not have proof reading activity.

Q. 347. Which describes a transcription unit?

(a) Strand of DNA between promoter and terminator **Answer** : (a)
(b) Strand of DNA between operator and terminator
(c) Strand of DNA between repressor and terminator
(d) Strand of DNA between enhancer and terminator.

Q. 348. RNA Polymerase of E. Coli requires which metal for its activity?

(a) Copper **Answer** : (b)
(b) Zinc
(c) Selenium
(d) Molybdenum.

Q. 349. α-Amanitin typically is used to inhibit which enzyme?

(a) Eukaryotic RNA polymerase II **Answer** : (a)
(b) Eukaryotic DNA polymerase I
(c) Eukaryotic DNA polymerase II
(d) Eukaryotic RNA polymerase I.

Q. 350. Which RNA polymerase is responsible for tRNA synthesis?

(a) RNA polymerase I **Answer** : (c)
(b) RNA polymerase II
(c) RNA polymerase III
(d) No specific RNA polymerase.

Q. 351. Usually which one of the following could serve as a consensus sequence for transcription termination in eukaryotes?

(a) AAUAAA **Answer** : (a)
(b) AGUAGG
(c) TTATTA
(d) GCGCGCGC.

Q. 352. Which codon acts as an initiator codon during protein synthesis?

(a) AUG (b) AUC **Answer** : (a)
(c) UGA (d) UAG.

Q. 353. Which amino acid has a single codon?

(a) Tryptophan **Answer** : (a)
(b) Lysine
(c) Leucine
(d) Valine.

Q. 354. Degeneracy of codons means what?

(a) Codons are degenerated **Answer :** (b)
(b) Multiple codons decode the same amino acid
(c) Codons of mRNA degenerate after one use during protein synthesis
(d) Defective codons.

Q. 355. Which statement is true?

(a) First nucleotide of a codon is less important than third codon **Answer :** (c)
(b) There are only two nonsense codons
(c) Some tRNA molecules can utilise the same anticodon to recognise more than one amino acid
(d) The third nucleotide of codon binds tightly with the third nucleotide of anticodon during translation.

Q. 356. In Hemoglobin S which is correct?

(a) Valine replaces glutamic acid at 6th position in β chain **Answer :** (a)
(b) Glutamic acid replaces valine at 6th position in β chain
(c) Valine replaces glutamic acid at 6th position in α-chain
(d) Glutamic acid replaces valine at 6th position in α-chain.

Q. 357. Which is the rate limiting step in protein synthesis?

(a) Recognition of mRNA cap by 4E protein **Answer :** (a)
(b) Formation of preintiation complex
(c) Assembly of 40S ribosome with 4F protein
(d) Formation of 80S ribsosome complex.

Q. 358. How many high energy phosphate bonds are hydrolysed for one peptide bond formation during protein synthesis?

(a) 2 **Answer :** (c)
(b) 3
(c) 4
(d) 1.

Q. 359. Usually how many amino acids are incorporated per second by eukaryotic ribosome?

(a) 4 **Answer :** (b)
(b) 6
(c) 8
(d) 10.

Q. 360. Which is not true?

(a) Many ribosomes can translate the same mRNA simultaneously **Answer :** (d)
(b) Single mammalian ribosome is capable of synthesizing about 400 peptide bonds per minute
(c) Ribosomes cannot attach to same mRNA any closer than 80 nucleotides apart
(d) In eukaryotes when the anticodon of tRNA recognises a termination codon protein synthesis stops.

Q. 361. Tetracycline prevents bacterial protein synthesis by

(a) Preventing binding of aminoacyl tRNA to A site — **Answer : (a)**
(b) Preventing assembly of 30S and 50S ribosome
(c) Inhibiting peptidyl transferase enzyme
(d) Inhibiting aminoacyl tRNA synthetase enzyme.

Q. 362. ADP ribosylation of eEF-2 is brought about by which one?

(a) Tetracycline — **Answer : (c)**
(b) Erythromycin
(c) Diphtheria toxin
(d) Puromycin.

Q. 363. Which one is the start codon for protein synthesis?

(a) UGA (b) CGA — **Answer : (c)**
(c) AUG (d) CUG.

Q. 364. Which of these is required during termination of protein synthesis and release of the peptide and tRNA from P site of ribosome?

(a) ATP (b) CTP — **Answer : (c)**
(c) GTP (d) UTP.

Q. 365. Translocation step of protein synthesis requires which one?

(a) No energy — **Answer : (b)**
(b) Hydrolysis of GTP to GDP
(c) Hydrolysis of ATP to ADP
(d) Hydrolysis of CTP to CDP.

Q. 366. Which one is an operon?

(a) An operator gene — **Answer : (c)**
(b) A gene controlled by a single promoter
(c) A linear array of genes regulated by a single promoter or regulatory region
(d) A constitutive gene regulated by single promoter.

Q. 367. Who described the operon model?

(a) Hamilton smith — **Answer : (b)**
(b) Jacob and Monod
(c) Messelstein and Stahl
(d) Avery and Macloed.

Q. 368. Which statement is not true?

(a) Acetylation of histones helps in transcription — **Answer : (c)**
(b) Acetylation of histone occurs on amino terminal lysine residues of histones
(c) Methylation of deoxy cytidine stimulates transcription
(d) Histone acetylase activity is associated with TAFs.

Q. 369. Which is not true for mRNA?

(a) mRNA exists in the cytoplasm as ribonucleoprotein particles (RNPs) — **Answer : (c)**
(b) There are noncoding sequences in 5' end
(c) Poly A tail prohibit the action of 5' exonuclease
(d) mRNAs may have half lives of 10.30 minutes.

Q. 370. Which is not true for lac operon?

(a) The genes for lactose metabolism follow a coordinate expression **Answer :** (d)

(b) Each of the enzyme is translated separately

(c) The mRNA of the three enzymes of lac operon is polycistronic in nature

(d) From a single large precursor protein, the three enzymes are cleaved.

Q. 371. To which carbon of ribose is the base attached in a nucleotide structure?

(a) C_1 (b) C_2 **Answer :** (a)

(c) C_3 (d) C_5.

Q. 372. What defines orientation of given strand of DNA?

(a) Adenine pairing with thymine **Answer :** (c)

(b) Cytosine pairing with guanine

(c) 3'-5' phosphodiester linkage

(d) Stacking of bases in a helical form.

Q. 373. Which is an incorrect statement?

(a) Coding regions are called exons **Answer :** (d)

(b) Non coding regions are called introns

(c) Several genes can be controlled by a single regulatory unit

(d) Introns tend to be always smaller than exons.

Q. 374. Restriction endonuclease enzymes are so named initially because of which one of the reasons given?

(a) They restricted the growth of bacteria **Answer :** (c)

(b) Conferred resistance to antibiotic among bacteria

(c) Restricted growth of bacteriophages in bacteria

(d) Restricted replication of bacterial DNA.

Q. 375. Which enzyme always stays is pair with restriction enzymes in bacteria?

(a) Site specific DNA methylases **Answer :** (a)

(b) DNA polymerases

(c) DNA directed RNA polymerases

(d) Site specific RNA methylases.

Q. 376. What confers antibiotic resistance to bacteria?

(a) Bacteriophage **Answer :** (c)

(b) Restriction enzymes

(c) Plasmids

(d) Lysosomes.

Q. 377. What is the action of the enzyme terminal transferase?

(a) Adds nucleotide at 5' end of DNA **Answer :** (b)

(b) Adds nucleotide of 3' end of DNA

(c) Transfer terminal group of nucleotides from one DNA to another DNA

(d) Transfer terminal group of nucleotides from DNA to RNA.

Q. 378. For DNA sequencing who first used dideoxynucleotides?

(a) F. Sanger (b) Maxam and Gilbert **Answer :** (a)
(c) Temin (d) H. Smith.

Q. 379. In PCR technique DNA polymerase obtained from which organism is used?

(a) Thermus aquaticus **Answer :** (a)
(b) H. influenze
(c) Drosophila
(d) E. Coli.

Q. 380. In which chromosome is insulin gene located in humans?

(a) Chromosome 11 **Answer :** (a)
(b) Chromosome 17
(c) Chromosome 6
(d) Chromosome 22.

Q. 381. Which one of these can be a palindrome?

(a) GATACC, CTATGG (b) GGATCC CCTAGG **Answer :** (b)
(c) TTACTTAAG, AATGATTC (d) CGAACGA, GCTTGCT.

Q. 382. Which one can degrade DNA both at 3' and 5' end?

(a) Alkaline phosphatase **Answer :** (b)
(b) BAL 31 nuclease
(c) Exonuclease III
(d) Polynucleotide kinase.

Q. 383. For production of homopolymer tailing in RNA which enzyme is required?

(a) Polynucleotide kinase **Answer :** (b)
(b) Terminal transferase
(c) DNA polymerase I
(d) ST nuclease.

Q. 384. Approximately what percentage of human genome encodes protein?

(a) 1%-1·5% **Answer :** (a)
(b) 2%-3%
(c) 10%-15%
(d) 20%-25%.

Q. 385. Which chromosome has richest gene content in humans as understood by human genome project?

(a) Chromosome 11 **Answer :** (b)
(b) Chromosome 19
(c) Chromosome 13
(d) Chromosome 6.

07 MEMBRANES AND HORMONES

Q. 386. Which is the major intracellular anion?

(a) Phosphate **Answer : (a)**
(b) Chloride
(c) Bicarbonate
(d) Lactate.

Q. 387. Which is the major extracellular anion?

(a) Phosphate **Answer : (b)**
(b) Chloride
(c) Bicarbonate
(d) Lactate.

Q. 388. Which is the major intracellular cation?

(a) Sodium **Answer : (b)**
(b) Potassium
(c) Calcium
(d) Magnesium.

Q. 389. Which is the major extracellular cation?

(a) Sodium (b) Potassium **Answer : (a)**
(c) Calcium (d) Magnesium.

Q. 390. Which enzyme acts as a marker for endoplasmic reticulum?

(a) 5 Nucleotidase **Answer : (b)**
(b) Glu-6-phosphatase
(c) Sialyl transferase
(d) Galactosyl transferase.

Q. 391. Which is not true of membrane structure?

(a) Choline containing phospholipids are located mainly in the outer layer **Answer : (c)**
(b) Phosphatidyl ethanolamine is usually found in the inner layer
(c) Cholesterol is generally present in large amounts on the inner side
(d) Unsaturated fatty acids provide fluidity to membrane structure.

Q. 392. Which is not true for membrane proteins?

(a) Membrane proteins are either integral or peripheral type **Answer : (b)**
(b) Integral proteins are symmetrically distributed across the membrane bilayer
(c) Peripheral proteins do not interact directly with the phospholipid bilayer
(d) Ankyrin is a peripheral protein.

Q. 393. Which is true?

(a) Immunoglobulins on the plasma membrane of lymphocytes are integral membrane proteins **Answer** : (a)
(b) Immunoglobulins on the plasma membrane of lymphocytes are peripheral protein
(c) Hormone receptors are always peripheral proteins
(d) Peripheral proteins require detergents for their release from plasma membranes.

Q. 394. Which is not correct?

(a) Fluidity of membranes depends on lipid composition of membrane **Answer** : (c)
(b) Cholesterol modifies fluidity of membranes
(c) Lateral mobility of integral protein decreases with increase in fluidity of membrane
(d) Increased fluidity of membrane alters the insulin receptor so that it binds more insulin.

Q. 395. Signal hypothesis for membrane proteins was proposed by whom?

(a) Nicolson & Singer **Answer** : (b)
(b) Blobel & Sabatini
(c) Messelstein & Stahl
(d) Watson & Crick.

Q. 396. Which is not a molecular chaperon?

(a) Calnexin **Answer** : (d)
(b) Calreticulin
(c) Protein disulphide isomerase
(d) Clathrin.

Q. 397. Which statement is incorrect?

(a) Facilitated diffusion may involve carrier protein **Answer** : (b)
(b) Facilitated diffusion acts only unidirectionally
(c) There is no covalent interaction between solute & carrier protein in facilitated diffusion
(d) Structurally similar competitive inhibitors block transport.

Q. 398. DNA enters a cell during transfection by which process?

(a) Pinocytosis **Answer** : (c)
(b) Endocytosis
(c) Endosmosis
(d) Endomitosis.

Q. 399. Which ion triggers exocytosis?

(a) Na^+ (b) K^+ **Answer** : (c)
(c) Ca^{+2} (d) Mn^{+2}.

Q. 400. Macromolecules avoid lysosomes after internalisation into a cell by forming which vesicle?

(a) Endocytic vesicle (b) Pinocytic vesicle **Answer** : (c)
(c) Receptosome (d) None of the above.

Q. 401. Which one of these utilise cAMP as the second messenger?

Answer : (b)

(a) Calcitriol
(b) Calcitonin
(c) Atrial natriuretic factor
(d) Insulin.

Q. 402. Which one of these utilise cGMP as the second messenger?

Answer : (d)

(a) Glucagon
(b) Somatostatin
(c) ADH
(d) Nitric oxide.

Q. 403. Which one uses calcium or phosphatidyl inositol or both as second messenger?

Answr : (c)

(a) TSH
(b) Retinoic acid
(c) Oxytocin
(d) Insulin like growth factor.

Q. 404. Which one stimulates adenylyl cyclase enzyme?

Answer : (a)

(a) FSH
(b) Acetylcholine
(c) Somatostatin
(d) Angiotensin II.

Q. 405. Which one inhibits adenylyl cyclase enzyme?

Answer : (d)

(a) Calcitonin
(b) Glucagon
(c) TSH
(d) Somatostatin.

Q. 406. ADP—ribosylation of α_S subunit of GTP dependent regulatory proteins (G-protein) is produced by which one?

Answer : (a)

(a) Cholera toxin
(b) Pertussis toxin
(c) Shigella toxins
(d) Hemophilus toxin.

Q. 407. ADP—ribosylation of α_i subunit of GTP dependent regulatory protein (G-protein) is produced by which one?

Answer : (b)

(a) Cholera toxin
(b) Pertussis toxin
(c) Shigella toxin
(d) Hemophilus toxin.

Q. 8. Which is not true of cAMP?

Answer : (c)

(a) It is produced from ATP by the action of adenylyl cyclas
(b) It is hydrolysed by phosphodiesterases
(c) Effect of cAMP is seen only in the extranuclear part, i.e. cytosol
(d) cAMP is also detected in extracellular fluid.

Q. 409. Which hormone uses cGMP as the second messenger?

(a) FSH **Answer** : (d)
(b) Prolactin
(c) ADH
(d) Atrial natriuretic factor.

Q. 410. How many Ca^{+2} binding sites are present in calmodulin?

(a) Two **Answer** : (c)
(b) Three
(c) Four
(d) Eight.

Q. 411. Which phospholipase enzyme catalyse hydrolysis of phosphatidyl inositol 4-5 bisphosphate to inositol triphosphate?

(a) Phospholipase A **Answer** : (c)
(b) Phospholipase B
(c) Phospholipase C
(d) Phospholipase D.

Q. 412. The JAK pathway is typically associated with which hormone receptor interaction?

(a) Growth hormone **Answer** : (a)
(b) Thyroxine
(c) Aldosterone
(d) Glucagon.

Q. 413. Which is not true for growth hormone?

(a) Growth hormone receptor is a member of cytokine—haematopoiten receptor superfamily **Answer** : (c)
(b) Growth related effects of growth hormone are primarily mediated by IGF-1
(c) Effects of growth hormone on lipid metabolism is mediated by IGF-1
(d) Growth hormone has many properties of prolactin.

Q. 414. Which is true for prolactin?

(a) Prolactin is secreted by basophilic cells of anterior pituitary **Answer** : (b)
(b) Tumors of prolactin secreting cells cause amenorrhoea
(c) Receptor of prolactin has no similarity with that of growth hormone
(d) Prolactin does not have structural similarity with growth hormone.

Q. 415. Which is not a glycoprotein hormone?

(a) LH **Answer** : (d)
(b) TSH
(c) Chorionic gonadotropin (CG)
(d) Prolactin.

Q. 416. Which one does not belong to the pro-opiomelanocortin (POMC) peptide family?

(a) ACTH **Answer** : (d)
(b) LPH
(c) MSH
(d) Melatonin.

Q. 417. Which of these statements is true for pituitary gland?

(a) Posterior pituitary gland synthesizes POMC peptide **Answer** : (b)
(b) Intermediary lobe is rudimentary in adult humans
(c) Anterior pituitary synthesizes ADH
(d) TSH belongs to POMC peptide family.

Q. 418. Which statement is not true for ACTH?

(a) ACTH has 39 amino acids **Answer** : (d)
(b) 24 amino terminal amino acids are required for full biological activity
(c) ACTH enhances conversion of cholesterol to pregnenolone
(d) Prolonged ACTH stimulation result in excessive glucocorticoid production only.

Q. 419. Which statement is incorrect for oxytocin?

(a) Stimulation of nipples releases oxytocin **Answer** : (b)
(b) Progesterone stimulates oxytocin release
(c) Part of oxytocin is proposed to be prolactin releasing factor
(d) Uterine distension is a secondary stimulus for oxytocin release.

Q. 420. Which statement is incorrect for ADH?

(a) ADH is released form posterior pituitary **Answer** : (c)
(b) Increased plasma osmolality is the primary stimulus for ADH release
(c) Epinephrine increase ADH release
(d) Ethanol decreases ADH release.

Q. 421. Which one is incorrect?

(a) There are two types of ADH or vasopressin receptors **Answer** : (b)
(b) Extra renal receptors of ADH are V_2 type
(c) cAMP is thought to mediate the effects of ADH on renal tubules
(d) ADH acts on distal convoluted tubules & collecting tubules, increasing their permeability to water.

Q. 422. When iodine supplies are adequate what is the ratio of $T_4 : T_3$ in Thyroid gland?

(a) 4 : 1 (b) 5 : 1 **Answer** : (d)
(c) 6 : 1 (d) 7 : 1.

Q. 423. What is the daily iodide requirement in the diet for an normal adult?

(a) 50—100 μg **Answer** : (c)
(b) 100—150 μg
(c) 150—200 μg
(d) 1—2 mg.

Q. 424. During thyroid hormone synthesis which step is called organification?

(a) Production of tyrosine from thyroglobulin **Answer** : (c)
(b) Conversion of I^- to I^+
(c) Iodination of tyrosine to produce MIT & DIT
(d) Coupling of two DIT molecules to produce T_4.

Q. 425. Which is not correct for thyroid hormones?

(a) T_3 & T_4 cause positive nitrogen balance **Answer** : (c)

(b) Thyroid hormones are required for conversion of tadpole into a frog

(c) T_3 reduce transcription of GH gene

(d) Half life of T_4 is four to five times that of T_3 in plasma.

Q. 426. Which is not true for calcium?

(a) Calcium exists as hydroxy apatite crystals in bone **Answer** : (b)

(b) Most of the calcium in bone is not freely exchangeable with ECF calcium

(c) About 6% of total calcium is complexed with citrate, phosphate & other anions

(d) Hypoalbuminemia results in decrease of total plasma calcium estimation.

Q. 427. Which statement is not correct?

(a) 1,25 $(OH)_2$—D_3 causes reduced production of PTH mRNA **Answer** : (c)

(b) Hypocalcemia increases PTH production

(c) Amino terminal segments of PTH are found in the circulation in large amounts

(d) Carboxy terminal residues of PTH are found in large amounts in circulation.

Q. 428. Which one promotes translocation of calcium against concentration gradient in intestinal cell membrane?

(a) 1, 25 $(OH)_2$—D_3 **Answer** : (a)

(b) PTH

(c) Calcitonin

(d) ACTH.

Q. 429. C1-hydroxylation of 1,25 $(OH)_2$—D_3 occurs in which organ?

(a) Liver (b) Kidney **Answer** : (b)

(c) Spleen (d) Intestine.

Q. 430. Which is not true for vitamin D?

(a) The biologically active form of Vitamin D is 1, 25 $(OH)_2$—D_3 **Answer** : (b)

(b) The precursor molecule of synthesis of vitamin D is 5-dihydro cholesterol

(c) C—25 hydroxylation occurs in liver

(d) C—1 hydroxylation occurs in kidney.

Q. 431. Which one inhibits renal 1 α-hydroxylase activity during 1, 25 $(OH)_2$ —D_3 synthesis?

(a) Hypocalcemia **Answer** : (c)

(b) Parathormone

(c) Calcitriol

(d) Hypophosphatemia.

Q. 432. How many carbon atoms are present in glucocorticoids?

(a) 21 (b) 23 **Answer** : (a)

(c) 25 (d) 27.

Q. 433. In which area of adrenal gland cortisol is synthesized?

Answer : (b)

(a) Zona glomerulosa
(b) Zona fasciculata
(c) Zona reticularis
(d) Medullary part.

Q. 434. In which layer of adrenal gland is aldosterone synthesized?

Answer : (c)

(a) Zona reticularis
(b) Zona fasciculata
(c) Zona glomerulosa
(d) In all the three layers, no specific area.

Q. 435. Cholesterol when acted upon by side chain cleavage enzyme of mitochondria produces which one?

Answer : (a)

(a) Pregnenolone
(b) Androstenedione
(c) Progesterone
(d) Epiandrosterone.

Q. 436. During steroid hormone synthesis, hydroxylation of which carbon is necessary for both glucocorticoid & mineralocorticoid activity?

Answer : (d)

(a) C_{17}
(b) C_{18}
(c) C_{19}
(d) C_{21}.

Q. 437. Which one is the most potent adrenal androgen?

Answer : (c)

(a) Dehydroepiandrosterone
(b) Androstenedione
(c) Testosterone
(d) 17-hydroxy pregnenolone.

Q. 438. Which one of these steroid hormones does not have a specific plasma transport protein?

Answer : (d)

(a) Cortisol
(b) Corticosterone
(c) Progesterone
(d) Aldosterone.

Q. 439. Which of the statements is incorrect?

Answer : (c)

(a) Angiotensinogen is made in liver
(b) Renin is produced in juxta glomerular cells of kidney
(c) Estrogen decreases angiotensinogen production in liver while glucocorticoids enhance production
(d) Production of angiotensinogen II from angiotensin I is not rate limiting.

Q. 440. Which factor stimulates renin release?

Answer : (d)

(a) Vasopressin
(b) Potassium
(c) Angiotensin II
(d) Prostaglandins.

Q. 441. Which is the major product of adrenal medulla?

(a) Epinephrine
(b) Norepinephrine
(c) Dopamine
(d) Acetylcholine.

Answer : (a)

Q. 442. Which is the rate limiting enzyme in catecholamine biosynthesis?

(a) Tyrosine hydroxylase
(b) DOPA decarboxylase
(c) Dopamine β hydroxylase
(d) Phenyl methanolamine N—methyl transferase (PNMT).

Answer : (a)

Q. 443. Dopa decarboxylase requires which coenzyme?

(a) Biotin
(b) Pyridoxal phosphate
(c) TPP
(d) Tetrahydrofolate.

Answer : (b)

Q. 444. Measurement of which compound in urine is extremely helpful to diagnose pheochromocytoma?

(a) MAO (Mono amine oxidase)
(b) VMA (Vanillyl mandelic acid)
(c) Homovanillic acid
(d) 3 methoxy tyramine.

Answer : (b)

Q. 445. Catecholamines stimulate lipolysis acting via which adrenergic receptor?

(a) α_1
(b) α_2
(c) β_1
(d) β_2.

Answer : (c)

Q. 446. Hormones that bind to which one of these adrenergic receptor & can inhibit adenylyl cyclase?

(a) α_1
(b) α_2
(c) β_1
(d) β_2.

Answer : (b)

Q. 447. Which cells of testes are responsible for most of the production of testosterone?

(a) Sertoli cells
(b) Leydig cells
(c) Germ cells
(d) Basement membrane of seminiferous tubules.

Answer : (b)

Q. 448. Which one is mostly responsible for the stimulus to produce testosterone in leydig cells?

(a) FSH
(a) LH
(c) ACTH
(d) TSH.

Answer : (b)

Q. 449. Which one is the most significant metabolite of testosterone exhibiting androgenic activity?

(a) 17 keto steroid **Answer : (b)**
(b) Dihydrotestosterone
(c) Androstenedione
(d) Etiocholanolone.

Q. 450. Which statement is incorrect?

(a) In pregnancy more estriol is produced **Answer : (c)**
(b) Estrogens are produced by aromatisation of androgens
(c) Aromatase enzyme produces estradiol from androstenedione
(d) Corpus luteum can produce progesterone & some estradiol.

Q. 451. Which is not true for oestrogen?

(a) Oestrogen stimulates production of progesterone receptor **Answer : (b)**
(b) Oestrogen decreases peripheral blood flow, thereby decrease heat loss
(c) Oestrogen causes proliferation of breast ducts
(d) Estradiol has anabolic effect on bone.

Q. 452. Which is the major hormone of luteal phase of menstruation in humans?

(a) Oestrogen (b) Progesterone **Answer : (b)**
(c) ICSH (d) FSH.

Q. 453. Which one is produced in largest amounts during pregnancy?

(a) Estradiol **Answer : (c)**
(b) Estrone
(c) Estriol
(d) Oestrogens decrease during pregnancy.

Q. 454. Which is not true for placenta?

(a) Placenta synthesizes progesterone **Answer : (c)**
(b) Fetal adrenal produces DHEA which is converted by placenta to estriol
(c) Placenta synthesizes cholesterol to maintain progesterone level during pregnancy
(d) Fetal adrenals depend on placental progesterone for cortisol synthesis (in the foetus).

Q. 455. Which is untrue of lactation?

(a) Prolactin increases synthesis of milk protein **Answer : (d)**
(b) Progesterone inhibts milk production & secretion in late pregnancy
(c) Suckling stimulate oxytocin secretion
(d) Prolactin levels rise rapidly in postpartum, not withstanding the effects of suckling.

Q. 456. Who proved the association of insulin (cells of langerhans) with diabetes mellitus?

(a) Von Mering (b) De Mayer **Answer : (d)**
(c) Sharpey—Schaffer (d) Banting & Best.

Q. 457. How many amino acids are present in A chain of insulin?

(a) 20 **Answer** : (b)
(b) 21
(c) 30
(d) 31.

Q. 458. Which is not correct?

(a) Human pancreas secretes 40-50 units of insulin daily in normal adults **Answer** : (c)
(b) Insulin secretion is an energy requiring process
(c) Maximal response of insulin secretion can be achieved at blood glucose level of 140 mg/dl
(d) Ca^{++} influx into β cells result in insulin secretion.

Q. 459. Which metabolic pathway is not stimulated by insulin?

(a) Process of glycolysis **Answer** : (b)
(b) Process of lipolysis
(c) Process of protein synthesis
(d) Process of glycogenesis.

Q. 460. What is the half life of insulin receptor?

(a) 7—12 hrs **Answer** : (a)
(b) 1—2 hr
(c) 24 hrs
(d) 24—28 hrs.

Q. 461. Activity of which one of the enzymes is decreased by insulin?

(a) Glycogen synthase **Answer** : (d)
(b) Pyruvate kinase
(c) HMG CoA reductase
(d) Triacylglycerol lipase.

Q. 462. Which hormone does not belong to gastrin family of gastrointestinal hormones?

(a) Gastrin **Answer** : (c)
(b) Cholecystokinin
(c) Secretin
(d) None of the above.

Q. 463. Pancreatic amylase secretion is stimulated by which one?

(a) Gastrin **Answer** : (c)
(b) Secretin
(c) Cholecystokinin
(d) Pancreatic polypeptide.

08 VITAMINS AND NUTRITION

Q. 464. Oxidative decarboxylation reaction requires coenzymic form of which vitamin?

(a) Thiamine
(b) Riboflavin
(c) Niacin
(d) Biotin.

Answer : (a)

Q. 465. Wernicke's encephalopathy is associated with which vitamin deficiency?

(a) Folic acid
(b) Riboflavin
(c) Thiamine
(d) Pantothenic acid.

Answer : (c)

Q. 466. Erythrocyte transketolase activity is used to measure deficiency of which vitamin?

(a) Thiamine
(b) Riboflavin
(c) Niacin
(d) Pyridoxine.

Answer : (a)

Q. 467. Erythrocyte glutathione reductase enzyme activity is used to assay which vitamin deficiency?

(a) Thiamine
(b) Riboflavin
(c) Pantothenic acid
(d) Biotin.

Answer : (b)

Q. 468. Which one of the these enzymes does not require FAD as coenzyme?

(a) Aldehyde dehydrogenase
(b) Succinate dehydrogenase
(c) Mitochondrial glycerol-3-phosphate dehydrogenase
(d) β-hydroxy acyl dehydrogenase.

Answer : (d)

Q. 469. Which vitamin is a pyridine derivative?

(a) Niacin (b) Thiamine
(c) Riboflavin (d) Folic acid.

Answer : (a)

Q. 470. Niacin can be generated from which amino acid?

(a) Tyrosine (b) Phenylalanine
(c) Tryptophan (d) Arginine.

Answer : (c)

Q. 471. Active form of which vitamin is involved in synthesis of niacin from tryptophan, such that its deficiency potentiate the deficiency of niacin?

(a) Pyridoxine
(b) Biotin
(c) Folic acid
(d) Vitamin B_{12}.

Answer : (a)

Q. 472. β-alanine is present in which vitamin?

(a) Folic acid
(b) Pantothenic acid
(c) Pyridoxin
(d) Biotin.

Answer : (b)

Q. 473. Transamination requires the coenzymic activity of which vitamin?

(a) Riboflavin
(b) Niacin
(c) Pantothenic acid
(d) Pyridoxine.

Answer : (d)

Q. 474. 70-80% of the total of which vitamin in the body is found in muscle phosphorylase enzyme?

(a) Thiamine
(b) Pyridoxine
(c) Folic acid
(d) Riboflavin.

Answer : (b)

Q. 475. Coenzymic form of which vitamin forms an integral part of phosphorylase, the enzyme catalysing breakdown of glycogen.

(a) Thiamine
(b) Biotin
(c) Pyridoxine
(d) Riboflavin.

Answer : (c)

Q. 476. Active form of which vitamin acts as a coenzyme for carboxylation reaction?

(a) Thiamine
(b) Pyridoxine
(c) Biotin
(d) Folic acid.

Answer : (c)

Q. 477. Avidin, a heat labile protein of egg white, can cause which vitamin deficiency?

(a) Biotin (b) Pantothenic acid
(c) Folic acid (d) Niacin.

Answer : (a)

Q. 478. Which coenzymic form of vitamin B_{12} is required for conversion of L—methylmalonic acid to succinyl CoA?

(a) Hydroxycobalamin
(b) Deoxy adenosyl cobalamin
(c) Methyl cobalamin
(d) Cobalamin.

Answer : (b)

Q. 479. Which coenzymic form of vitamin B_{12} is required to generate methionine from homocysteine?

(a) Hydroxy cobalamin **Answer** : (c)
(b) 5 deoxy adenosyl cobalamin
(c) Methyl cobalamin
(d) Trans cobalamin.

Q. 480. Which amino acid is the major source of one carbon unit in the form of methylene group carried by tetrahydro fotale?

(a) Threonine **Answer** : (b)
(b) Serine
(c) Methionine
(d) Glycine.

Q. 481. Neural tube defect like spina bifida can be prevented in foetuses by supplementing which vitamin in the diet of mothers in periconceptual period?

(a) Folic acid **Answer** : (a)
(b) Thiamine
(c) Vit B_{12}
(d) Riboflavin.

Q. 482. Deficiency of L-gulonolactone oxidase is responsible for non-synthesis of which vitamin in humans?

(a) Vitamin C **Answer** : (a)
(b) Folic acid
(c) Menadione
(d) Nicotinic acid.

Q. 483. Which one of the steps does not require active participation of vitamin C?

(a) Hydroxylation of proline & lysine during collagen synthesis **Answer** : (d)
(b) 7-α hydroxylation step of bile acid synthesis
(c) Dopamine β-hydroxylase step during epinephrine synthesis from tyrosine
(d) Conversion of Acetyl CoA to malonyl CoA.

Q. 484. Which one of these vitamins is a polyisoprenoid compound?

(a) Retinol **Answer** : (a)
(b) Folic acid
(c) Cholecalciferol
(d) Pyridoxal phosphate.

Q. 485. Oxidative cleavage of one molecule of β-carotene by β-carotene dioxygenase generates how many molecules of retinal?

(a) Two **Answer** : (a)
(b) Three
(c) Four
(d) Eight.

Q. 486. Which one has been implicated in controlling lungs surfactant phospholipid synthesis?

(a) Retinol **Answer : (c)**
(b) Retinal
(c) Retinoic acid
(d) β-carotene.

Q. 487. Which is the major carrier of β-carotene?

(a) HDL **Answer : (b)**
(b) LDL
(c) Albumin
(d) Retinol binding protein.

Q. 488. Which is the major form of vitamin D in circulation in humans?

(a) 25—hydroxy vitamin D_3 **Answer : (a)**
(b) 24, 25—dihydroxy vitamin D_3
(c) 1,25—dihydroxy vitamin D_3
(d) 25, 26—dihydroxy vitamin D_3.

Q. 489. Which is not true for vitamin E?

(a) D-α tocopherol has greatest biologic activity **Answer : (c)**
(b) Vitamin E appears to be the first line of defence against peroxidation of PUFA
(c) Antioxidant effect of vitamin E is effective at low oxygen concentration
(d) Vitamin E & selenium act synergistically.

Q. 490. Which is the major form of vitamin K found in plants?

(a) Phylloquinone **Answer : (a)**
(b) Menadione
(c) Napthoquinone
(d) Menaquinone.

Q. 491. What is the commonest cause of vitamin K deficiency?

(a) Deficiency of synthesis of clotting factors **Answer : (b)**
(b) Malabsorption of fat
(c) Deficiency of intake of vitamin K
(d) Vegeterian dietary habit.

Q. 492. Which of the group of clotting factors is dependent on vitamin K for their final active form?

(a) Factors II, VII, IX, X **Answer : (a)**
(b) Factors I, II, IX, XI
(c) Factors II, VII, VIII, X
(d) Factors I, VII, VIII, X.

Q. 493. How much of the total energy requirement of an healthy adult be supplied by essential fatty acids?

(a) 1—2% (b) 5—10% **Answer : (a)**
(c) 10—20% (d) 20—30%.

Q. 494. Which trace element is required for xanthine oxidase enzyme?

(a) Cobalt **Answer :** (b)
(b) Molybdenum
(c) Chromium
(d) Zinc.

Q. 495. Carbonic anhydrase, LDH & Alkaline phosphatase require which metal as a cofactor?

(a) Zinc **Answer :** (a)
(b) Selenium
(c) Molybdenum
(d) Chromium.

Q. 496. Glutathione peroxidase requires which trace element?

(a) Chromium **Answer :** (c)
(b) Zinc
(c) Selenium
(d) Nickel.

Q. 497. Which statement is not true?

(a) pH of gastric juice is about 1—1·5 **Answer :** (d)
(b) Parietal cells secrete intrinsic factor
(c) Pepsinogen is produced from chief cells
(d) Pepsin hydrolyses peptide bonds formed by basic amino acids.

Q. 498. Trypsin hydrolyzes peptide bonds formed by.

(a) Basic amino acids **Answer :** (a)
(b) Acidic amino acids
(c) Neutral amino acids
(d) Aromatic amino acids.

Q. 499. Which is the major end product of fat digestion?

(a) 1—monoacylglycerol **Answer :** (b)
(b) 2—monoacylglycerol
(c) Fatty acid
(d) 3—monoacylglycerol.

Q. 500. What is the end product of complete hydrolysis of Trehalose?

(a) Glucose only **Answer :** (a)
(b) Glucose & fructose
(c) Glucose, galactose & fructose
(d) Trehalose cannot be hydrolysed ever.

09 COLLAGEN, EXTRA-CELLULAR MATRIX AND MUSCLE

Q. 501. Digested products of which one of these is not absorbed into portal circulation?

(a) Fat **Answer : (a)**
(b) Protein
(c) Carbohydrate
(d) Minerals.

Q. 502. Which one is not a glycoprotein?

(a) Transferrin **Answer : (d)**
(b) Immunoglobulins
(c) Alkaline phosphatase
(d) ACTH.

Q. 503. Commonly how many types of monosaccharides are present in Glycoprotein?

(a) 4 **Answer : (c)**
(b) 6
(c) 8
(d) 5.

Q. 504. What is fucose?

(a) Hexose **Answer : (b)**
(b) Deoxyhexose
(c) Deoxy heptose
(d) Methyl hexose.

Q. 505. Which statement is incorrect?

(a) Lectins are sugar binding proteins & can agglutinate cells **Answer : (c)**
(b) Enzymes can be classed as lectins if they bind to carbohydrate
(c) Protein with single carbohydrate binding site will agglutinate cells
(d) Immunoglobulins that react with sugars are not considered lectins.

Q. 506. Dolichol is composed of which one?

(a) 17—20 repeating isoprenoid units **Answer : (a)**
(b) Repeating units of glucose & mannose
(c) Repeating units of N-acetyl glucosamine with mannose
(d) Repeating units of dulcitol with GlcNAc.

Q. 507. How may distinct types of collagen have been identified in human tissues?

(a) 10 (b) 14 **Answer** : (d)
(c) 17 (d) 19.

Q. 508. Which amino acid occurs at every third position of the triple helical portion of the α chains of collagens?

(a) Serine (b) Glycine **Answer** : (b)
(c) Proline (d) Lysine.

Q. 509. Collagen with discontinuous triple helices which form mesh-like network structure in basement membrane belongs to which type of collagen?

(a) Type I (b) Type II **Answer** : (c)
(c) Type IV (d) Type X.

Q. 510. In which of these conditions, there may be excessive production of collagen?

(a) Hepatic cirrhosis **Answer** : (a)
(b) Inflammatory states
(c) Starvation
(d) None of the above.

Q. 511. Which of the statement is true for elastin?

(a) Elastin has repeat (Gly-X-Y) sequence **Answer** : (c)
(b) It has triple helical structure
(c) It does not have carbohydrate moeity
(d) It has hydroxylysine & hydroxyproline both.

Q. 512. Which is not true for elastin?

(a) In nearly 90% cases of William's syndrome, there is deletion of elastin gene **Answer** : (b)
(b) Scleroderma is associated with depletion of elastin
(c) There may be decrease or fragmentation of elastin in pulmonary emphysema
(d) Lysine derived aldehydes in elastin help in cross link formation.

Q. 513. Which is not a correct statement?

(a) Fibrillin is a glycoprotein **Answer** : (c)
(b) Fibrillin is secreted by fibroblasts
(c) Fibrillin is not found in periosteum
(d) Gene for fibrillin is found on chromosome 15.

Q. 514. Which statement is true?

(a) Amount of carbohydrate in proteoglycans is usually much greater than is found in glycoproteins **Answer** : (a)
(b) Glycoproteins have more carbohydrate than that is usually present in proteoglycans
(c) Aggrecan is a glycoprotein of collagen
(d) Glycosaminoglycans are not found in proteoglycans.

Q. 515. Which glycosaminoglycan has a repeating disaccharide unit containing GIcUA & GIcNAC?

(a) Hyaluronic acid
(b) Chondroitin sulphate
(c) Heparin
(d) Dermatan sculphate.

Answer : (a)

Q. 516. Which one of these glycosaminoglycans is not sulphated?

(a) Hyaluronic acid (b) Heparin
(c) Heparan (d) Keratan sulphate.

Answer : (a)

Q. 517. About 90% of total bone protein belongs to which type of collagen?

(a) Collagen type I
(b) Collagen type II
(c) Collagen type III
(d) Collagen type IV.

Answer : (a)

Q. 518. Which one of these proteins is not bone specific?

(a) Osteocalcin
(b) Osteopontin
(c) Bone sialoprotein
(d) Chondroitin—proteoglycan (CS-PG III).

Answer : (b)

Q. 519. Which one inhibits osteoblasts?

(a) Thyroxine
(b) Parathormone
(c) Corticosteroids
(d) 1,25 dihydroxy cholecalciferol.

Answer : (c)

Q. 520. Which can stimulate both osteoblasts & osteoclasts?

(a) Parathormone (b) Calcitonin
(c) PGE_2 (d) Corticosteroid.

Answer : (a)

Q. 521. Which one can inhibit osteoclasts?

(a) Calcitonin
(b) Parathormone
(c) 1,25 dihydroxy cholecalciferol
(d) Interleukin—1.

Answer : (a)

Q. 522. Which is the major protein in hyaline cartilage?

(a) Collagen type II
(b) Collagen type I
(c) Collagen type V
(d) Collagen type IV.

Answer : (a)

Q. 523. Which statement is not true of Achondroplasia?

(a) Achondroplasia is the commonest cause of short limbed dwarfism
(b) It may be inherited as autosomal dominant trait
(c) It is basically a collagen disorder
(d) Fibroblast growth factor receptor 3 (FGFR 3) is defective or inactive in achondroplasia.

Answer : (c)

Q. 524. Which is functional unit of muscle?

Answer : (c)

(a) Sarcoplasm
(b) Sarcolemma
(c) Sarcomere
(d) Myofibril.

Q. 525. Between which two lines (bands) is sarcomere present?

Answer : (b)

(a) Between two A lines
(b) Between two Z lines
(c) Between two I lines
(d) Between one A & one Z line.

Q. 526. What is not true for muscle contraction?

Answer : (d)

(a) During muscle contraction H zones & I bands shorten
(b) There no change in lengths of thick & thin filaments
(c) Tension developed during muscle contractions is proportional to filament overlap
(d) Tension developed during muscle contraction is not dependent on cross links between thick & thin filament.

Q. 527. Which forms thick filament?

Answer : (b)

(a) Actin
(b) Myosin
(c) Troponin
(d) Tropomyosin.

Q. 528. During the cyclic process of muscular contraction & relaxation which event occurs first?

Answer : (c)

(a) Formation of Actin + Myosin
(b) Formation of Actin + Myosin + ATP
(c) Formation of Myosin + ATP
(d) Formation of Actin + ATP.

Q. 529. In straited muscles which is a calcium binding polypeptide/protein?

Answer : (a)

(a) Troponin C
(b) Troponin I
(c) Troponin T
(d) Tropomyosin.

Q. 530. How many molecules of calcium ions are bound per molecule of troponin C?

Answer : (b)

(a) Two
(b) Four
(c) Eight
(d) One.

Q. 531. In striated muscle which one blocks or inhibits muscle contraction?

Answer : (b)

(a) Troponin C
(b) Troponin I
(c) Troponin T
(d) F—actin of thin filament.

Q. 532. Malignant hyperthermia could be due to sudden high cytosolic concentration of Ca ions in skeletal muscles. Which drug is useful in this condition?

(a) Dantrolene
(b) Ryanodine
(c) Calsequestrin
(d) Succinylcholine.

Answer : (a)

Q. 533. In humans which is stated to be the largest known protein?

(a) Actin
(b) Myosin
(c) Titin
(d) Nebulin.

Answer : (c)

Q. 534. Which of the muscles lack troponin system?

(a) Striated (skeletal) muscles
(b) Cardiac muscles
(c) Smooth muscles
(d) All have troponin system.

Answer : (c)

Q. 535. NO synthase enzymes act on arginine to produce nitric oxide & which one of these compounds?

(a) Ornithine
(b) Citrulline
(c) Arginosuccinate
(d) Aspartate.

Answer : (b)

Q. 536. Nitric oxide can be inhibited by which one?

(a) Hemoglobin
(b) cGMP
(c) Calcium ions
(d) Inositol triphosphate.

Answer : (a)

Q. 537. Fatigue of muscle is produced due to which one of the mechanisms?

(a) Production of lactate
(b) Production of protons
(c) Increase release of Ca^{+2}
(d) Bicarbonate ions.

Answer : (b)

Q. 538. Urinary output of which compound is a reliable index of the rate of myofibrillar protein breakdown in the musculature of humans?

(a) Methyl histidine
(b) Glycine
(c) Methyl valine
(d) Methyl adenosine.

Answer : (a)

10 PLASMA PROTEINS AND BLOOD

Q. 539. Approximately what is the half life of plasma albumin?

(a) 5 days (b) 10 days

(c) 20 days (d) 30 days.

Answer : (c)

Q. 540. Which one of the statements is incorrect?

(a) Hemopexin can bind free heme

(b) Haptoglobin binds extracorpuscular hemoglobin

(c) Albumin can bind to plasma tryptophan

(d) Haptoglobin when bound to hemoglobin clears from plasma slowly than normal.

Answer : (d)

Q. 541. In which component, maximum amount of iron is distributed out of the total iron content in an healthy male?

(a) Hemoglobin

(b) Storage form as ferritin & hemosiderin

(c) Transferrin

(d) Myoglobin.

Answer : (a)

Q. 542. In Wilson's disease, which is untrue?

(a) Ceruloplasmin levels are low

(b) Ceruloplasmin level are high

(c) There is hepatolenticular degeneration

(d) There is copper toxicosis.

Answer : (b)

Q. 543. Which is a serine protease inhibitor?

(a) α_1-Antitrypsin

(b) Ceruloplasmin

(c) Trypsin

(d) Elastase.

Answer : (a)

Q. 544. Which is not true for α_2-macroglobulin?

(a) It carries 10% of plasma zinc

(b) It is a panprotease inhibitor

(c) It comprises 20% of the total plasma protein in humans

(d) It is a member of the complement proteins C_3 & C_4.

Answer : (c)

Q. 545. Which part of immunoglobulin molecule binds to specific antigen?

(a) Variable part of H—chain

(b) Variable part of L—chain

(c) Variable part of both H & L chains

(d) Constant part of H. chain.

Answer : (c)

Q. 546. One molecule of IgG can bind how many molecules of antigen?

(a) One (b) Two **Answer** : (b)
(c) Four (d) Eight.

Q. 547. Percentage of which immunogloblin is the least, out of the total immunoglobulins present in serum?

(a) IgG (b) IgD **Answer** : (d)
(c) IgA (d) IgE.

Q. 548. Who developed the technique of monoclonal antibody formation?

(a) F. Sanger **Answer** : (b)
(b) Kohler & Milstein
(c) Goldstein & Brown
(d) Hamilton Smith.

Q. 549. During production of hybridomas which one helps in fusion of B cells with myeloma cells?

(a) Polyethylene glycol **Answer** : (a)
(b) Hypoxanthine
(c) Aminopterine
(d) Thymidine.

Q. 550. Which is not true of thrombins?

(a) White thrombins are rich in platelets & fibrin **Answer** : (d)
(b) Red thrombins are rich in RBC & fibrin
(c) Disseminated fibrin deposit in capillaries
(d) White thrombins form particularly in places with retarded blood flow.

Q. 551. Which one of the clotting factors is not associated with intrinsic pathway of blood coagulation?

(a) Factor VII **Answer** : (a)
(b) Factor XII
(c) Factor XI
(d) Factor IX.

Q. 552. Tissue factor pathway inhibitor (TFPI) is a major physiologic inihibitor of coagulation. Which one of these is inhibited by it directly?

(a) Factor XIIa **Answer** : (c)
(b) Factor XIa
(c) Factor Xa
(d) Factor IXa.

Q. 553. The activity of anti-thrombin III (an inhibitor of thrombin) is increased by which one?

(a) α_2-Macroglobulin **Answer** : (b)
(b) Heparin
(c) α_1-Antitrypsin
(d) Protamin.

Q. 554. Which factor is involved in adherence of platelets to collagen of subendothelium?

(a) Von Willebrand factor **Answer : (a)**
(b) Anti hemophilic globulin B
(c) Plasminogen
(d) Factor V Leiden.

Q. 555. Which test is a measure of coagulation by intrinsic pathway mechanisms?

(a) Activated partial thromboplastin time (APTT) **Answer : (a)**
(b) Prothrombin time (PT)
(c) Thrombin time (TT)
(d) Platelet count.

Q. 556. Which test is a measuse of coagulation by extrinsic pathway mechanisms?

(a) APTT **Answer : (b)**
(b) PT
(c) TT
(d) Platelet count.

Q. 557. Effectiveness of oral anticoagulants like warfarin is assessed by measuring which one?

(a) APTT **Answer : (b)**
(b) PT
(c) TT
(d) Fibrinogen degradation product.

Q. 558. Which test is usually performed to monitor heparin therapy?

(a) APTT (b) PT **Answer : (a)**
(c) TT (d) Fibrinogen content.

Q. 559. Approximately what percentage of hemoglobin may auto-oxidise per day to methemoglobin?

(a) Nil (b) 1% **Answer : (c)**
(c) 3% (d) 7%.

Q. 560. Which does not act as an antioxidant?

(a) Vitamin C **Answer : (c)**
(b) Vitamin E
(c) H_2O_2
(d) Reduced Glutathione.

Q. 561. Which is not true of methemoglobinemia?

(a) Methemoglobinemia may be inherited or acqured due to ingestion of drugs **Answer : (c)**
(b) It is treated by administering methylene blue
(c) Cyanosis may be produced when methemoglobin level is 5%
(d) Inherited form of methemoglobinemia is usually due to deficiency of methemoglobin reductase enzyme.

Q. 562. Which is not a red cell membrane protein?

(a) Spectrin
(b) Actin
(c) Tropomysin
(d) Desmin.

Answer : (d)

Q. 563. Which of the group is the major integral membrane protein of RBC?

(a) Anion exchange protein & Glycophorins
(b) Spectrin & Ankyrin
(c) Glycophrin & Ankyrin
(d) Spectrin & Glycophorin.

Answer : (a)

Q. 564. Which group is the peripheral membrane proteins of RBC?

(a) Spectrin & Ankyrin
(b) Glycophorin & Actin
(c) Actin & Anion exchange protein
(d) Glycophorin & Spectrin.

Answer : (a)

Q. 565. Which is true for spherocytosis?

(a) Spherocytes have high surface to volume ratio
(b) Spherocytes have low surface to volume ratio
(c) Spherocytic anemia is a X-linked disease
(d) Nearly 100% spherocytes are lysed in 0·5 g/dl NaCl soln.

Answer : (b)

Q. 566. In which chromosome the genes of ABO blood group substance is present?

(a) Chromosome 6
(b) Chromosome 9
(c) Chromosome 11
(d) Chromosome 16.

Answer : (b)

Q. 567. Which protein mediate adhesion of neutrophils to endothelial cells?

(a) Ankyrin
(b) Integrin
(c) Vinculin
(d) Actin.

Answer : (b)

Q. 568. Presence of which enzyme in neutrophils give green colour of pus?

(a) NADPH oxidase
(b) Myeloperoxidase
(c) Cathepsin G
(d) Lysosomal enzymes.

Answer : (b)

11 XENOBIOTICS, GROWTH FACTORS AND CANCER

Q. 569. Cytochrome P_{450} is so named because of which reason?

(a) It has 450 amino acids **Answer : (d)**
(b) It has 450 varieties
(c) One molecule can maximum catalyse 450 molecules
(d) Exhibits peak optical density of 450 nm.

Q. 570. In which cell organelle is cytochrome P_{450} mostly found?

(a) Golgi complex **Answer : (d)**
(b) Mitochondria
(c) Free in cytosol
(d) Membranes of smooth endoplasmic recticulum.

Q. 571. Epoxides may be produced by action of monoxygenases. Since epoxides are mutagenic they are subject to conversion by an enzyme called epoxide hydrolase. What product is formed when epoxide hydrolase acts on epoxides?

(a) Water **Answer : (c)**
(b) Hydrodiol
(c) Dihydrodiol
(d) Hydrogen peroxide.

Q. 572. Glutathione is composed of which amino acids?

(a) Glutamic acid, cysteine, glycine **Answer : (a)**
(b) Arginine, glycine, glutamic acid
(c) Methionine, glycine, glutamic acid
(d) Valine, lysine, cysteine.

Q. 573. Mercapturic acid produced during detoxification reaction using glutathione is water soluble & excreted in urine. It is a conjugate product of which compound?

(a) L-acetyl cysteine **Answer : (a)**
(b) L-cysteinyl glycine
(c) L-glutamyl cysteine
(d) L-acetyl glycine.

Q. 574. Which may not be a function of glutathione?

(a) Decomposition of hydrogen peroxide **Answer : (c)**
(b) Transport of amino acid in kidney
(c) Intracellular oxidant
(d) Conjugation reaction during detoxification.

Q. 575. Spontaneous mutations some of which can predispose to cancer may occur at what frequency?

(a) 10^{-3} per cell per generation **Answer :** (d)
(b) 10^{-4} per cell per generation
(c) 10^{-5} per cell per generation
(d) 10^{-6} per cell per generation.

Q. 576. As a generalised statement, ultimate carcinogens are usually of which class?

(a) Electrophiles (b) Nucleophiles **Answer :** (a)
(c) Electrophobic (d) Nucleophobic.

Q. 577. Which base of DNA is most common site of attack by chemical carcinogens?

(a) Guanine (b) Adenine **Answer :** (a)
(c) Cytosine (d) Thymine.

Q. 578. Which assay is commonly employed to screen mutagenecity of potential carcinogens?

(a) Measurement of mercapturic acid **Answer :** (b)
(b) Ames assay
(c) Quantitation of cytochrome P_{450}
(d) Chromatogram of carcinogen.

Q. 579. Which is stated to be the product of 'Src' gene of Rous sarcoma virus which may bring about transformation of a cell?

(a) Reverse transcriptase **Answer :** (b)
(b) Protein tyrosine kinase
(c) Protein C kinase
(d) Vinculin protein.

Q. 580. Regarding chromosomal translocation phenomena which is not correct?

(a) Philadelphia chromosome is an example of translocation involving chromose 22 & 9 **Answer :** (c)
(b) In Burkitt's lymphoma translocation involves chromosome 8 & 14
(c) In Burkitt's lymphoma segment of chromosome 14 breaks off & joins with chromosome 8 containing c-myc
(d) In Philadelphia chromosome, there is production of bcr-bcl fusion protein.

Q. 581. Which is not correct about oncogenes?

(a) They may produce protein tyrosine kinase **Answer :** (d)
(b) They may produce products to stimulate adenylyl cyclase
(c) They may produce proteins to act as DNA binding proteins
(d) They are thought to be the sole pathway leading to carcinogenesis.

Q. 582. Platelet derived growth factor (PDGP) is released from which one?

(a) α-granules of platelets **Answer : (a)**
(b) β-granules of platelets
(c) γ-granules of platelets
(d) Golgi complex of platelets.

Q. 583. Which enzyme hydrolyses phosphatidyl inositol 4-5 diphosphate to form inositol triphosphate & diacyl glycerol?

(a) Phospholipase A_1 **Answer : (d)**
(b) Phospholipase A_2
(c) Phospholipase B
(d) Phospholipase C.

Q. 584. Which is not true of protein retinoblastoma (pRB)?

(a) pRB may regulate cell proliferation by binding to E2F transcription complex **Answer : (b)**
(b) pRB is a cytosolic phosphoprotein whose phosphorylation oscillates during cell cycle
(c) Phosphorylated form of pRB cannot bind to E2F
(d) Unphosphorylated form of pRB binds to certain vital proteins & may inactivate them.

Q. 585. Which gene is popularly called 'guardian of the genome' or 'molecular policeman'?

(a) p21 gene (b) p23 gene **Answer : (c)**
(c) p53 gene (d) p27 gene.

Q. 586. Which group of enzymes (proteases) are key factors for events of Apoptosis?

(a) Caspases **Answer : (a)**
(b) Product of bcl—2 gene
(c) Product of myc gene
(d) pRB.

Q. 587. Defect in mismatch repair system of DNA has been implicated in development of which cancer?

(a) Hereditary nonpolyposis colon cancer **Answer : (a)**
(b) Xeroderma pigmentosum
(c) Retinoblastoma
(d) Burkitt's lymphoma.

Q. 588. Defective nucleotide excision repair has been implicated in development of which disease?

(a) Colonic cancer (HNPCC) **Answer : (b)**
(b) Xeroderma pigmentosum
(c) Cancer breast
(d) Retinoblastoma.

Q. 589. Which of these confer multidrug resistance to cancer cells?

(a) P—Glycoprotein (b) p53 gene product **Answer : (a)**
(c) p21 gene product (d) Caspase protein.

Q. 590. Carcino embryonic antigen may be an useful tumour maker of cancer of which organ?

(a) Germ cell
(b) Prostate
(c) Colon
(d) Thyroid.

Answer : (c)

Q. 591. Caspases are associated with which cellular process?

(a) Cell differentiation
(b) Metastasis
(c) Apoptosis
(d) Mitosis.

Answer : (c)

Q. 592. Which disease may result due to mutations of mitochondrial DNA?

(a) Leber hereditary optic neuropathy
(b) Zellweger syndrome
(c) I—cell disease
(d) Sickle cell anemia.

Answer : (a)

Q. 593. I—cell disease is due to dysfunction of which major intracellular organelle?

(a) Nucleus
(b) Mitochondria
(c) Golgi complex
(d) Peroxisome.

Answer : (c)

Q. 594. Inborn error of which pathway is virtually unknown?

(a) Glycolysis
(b) HMP shunt pathway
(c) TCA cycle
(d) Urea cycle.

Answer : (c)

Q. 595. Lod score is used to determine which one?

(a) Linkage
(b) Disequilibrium
(c) Mutation
(d) Carcinogenecity.

Answer : (a)

Q. 596. Mutations of gene of which one of these is considered to be the cause of blindness in retinitis pigmentosa?

(a) Iris protein
(b) Rhodopsin
(c) Na^+ ion channel proteins
(d) Macular protein.

Answer : (b)

Q. 597. Which is presently the most powerful method (probably method of choice) to identify disease genes?

(a) Functional approach
(b) Candidate gene approach
(c) Positional cloning approach
(d) Postional candidate approach.

Answer : (d)

Q. 598. Among the choices given which has the smallest genome & least no of genes?

(a) Escherichia coli **Answer :** (d)
(b) Hemophilus influenzae
(c) Helicobacter pylori
(d) Trepenoma pallidum.

Q. 599. What is popularly called antisense compound?

(a) An oligonucleotide binding to mRNA or double stand DNA **Answer :** (a)
(b) A mRNA strand
(c) An artificial DNA complimentary to sense strand for DNA study
(d) A DNA strand complimentary to anticodons of tRNA.

Q. 600. Shotgun approach of genomic sequence analysis was initially used by researchers in the team led by whom?

(a) Craig Venter **Answer :** (a)
(b) Francis Collins
(c) James Watson
(d) F. sanger.

Q. 601. Which one of the vectors can take up the longest DNA by a recombinant technology?

(a) Phage **Answer :** (d)
(b) Plasmid
(c) Cosmid
(d) YAC.

12 NEUROPSYCHIATRIC DISORDERS

Q. 602. Which is not true for myasthenia gravis?

(a) It is a disease characterised by recurrent episodes of muscle weakness & fatigability

(b) There are autoantibodies to acetyl choline receptor of neuro muscular junction

(c) There is marked reduction in activity of the receptor due to inhibition rather than destruction of the receptor due to autoantibodies

(d) Muscles supplied by cranial nerves are most affected.

Answer : (c)

Q. 603. Which is not correct for acetylcholine receptor?

(a) It is a nicotinic receptor

(b) It is a membrane glycoprotein

(c) Only β subunits binds acetylcholine

(d) The snake venom toxin α-bungarotoxin binds to the receptor.

Answer : (c)

Q. 604. Acetylcholine receptor is which type of ion channel?

(a) Transmitter gated ion channel

(b) Voltage gated ion channel

(c) Mechanically gated ion channel

(d) None of the above.

Answer : (a)

Q. 605. In Huntington's disease neurons of which part of brain affected most?

(a) Neurons of corpus striatum

(b) Neurons of thalamus

(c) Neurons of cerebellum

(d) Neurons of temporal lobe.

Answer : (a)

Q. 606. Trinucleotide repeat expansion coding for which amino acid in "huntingtin gene" is believed to cause severity in huntington's disease?

(a) Glutamic acid (b) Serine

(c) Aspartic acid (d) Phenylalanine.

Answer : (a)

Q. 607. Nearly 75% of excitatory neurotransmission in the brain is dependent on which amino acid?

(a) Aspartic acid

(b) Glutamic acid

(c) Dopamine

(d) Glycine.

Answer : (b)

Q. 608. Fragile X syndrome is so named because of which one?

Answer : (d)

(a) It is X-linked with increased fragility of bones
(b) It is X-linked with fragile health
(c) It is X-linked & different chromosomes are fragile
(d) The X-chromosome has fragile sites.

Q. 609. Which one of the trinucleotide repeat expansion can cause chronic neurodegeneration diseases?

Answer : (a)

(a) CAG repeat (b) GGA repeat
(c) AAC repeat (d) UAU repeat.

Q. 610. How much of dopamine level is to be reduced in nigrostriatal system to produce symptoms of Parkinson's disease?

Answer : (d)

(a) 20% (b) 40%
(c) 60% (d) 80%.

Q. 611. Amyloid proteins derived their name because which of the choices given?

Answer : (b)

(a) They have structural similarity with amylose
(b) Stain blue colour with iodine
(c) Have amylopectin as a component
(d) Have both amylose & amylopectin components.

Q. 612. In myocardial infarction which one coincides with the development of irreversible cellular damage?

Answer : (c)

(a) Increased intracellular osmotic pressure
(b) Accumulation of lactic acid
(c) Exhaustion of adenine nucleotide pool
(d) Accumulation of protons.

Q. 613. Which reaction is catalyzed by superoxide dismutase?

Answer : (a)

(a) $O_2^{\bullet -} + O_2^{-} + 2H^+ \rightarrow H_2O_2 + O_2$
(b) $OH^\bullet + OH^\bullet \rightarrow H_2O_2$
(c) $O_2^{\bullet -} - e^- \rightarrow O_2$
(d) $O_2^{\bullet -} + e^- + 2H^+ \rightarrow H_2O$.

Q. 614. Which is not true for ethanol?

Answer : (c)

(a) It forms acetaldehyde which is further converted to acetate
(b) There is increase in lactate synthesis
(c) There is increased gluconeogenesis
(d) There is inhibition of fatty acid oxidation.

Q. 615. What is the diagnostic hallmark of cystic fibrosis?

Answer : (a)

(a) Presence of high amounts of NaCl in sweat
(b) Presence of decreased amout of NaCl in sweat
(c) Bronchopneumonia
(d) Diminished fat digestion.

Q. 616. Which is the first disease to be treated by somatic cell gene therapy?

(a) Duchenne type muscular dystrophy **Answer :** (b)
(b) ADA deficiency
(c) Cystic fibrosis
(d) Huntington's chorea.

Q. 617. Which one of the following is a measure of sensitivity of a test?

(a) $\frac{\text{True positive}}{\text{True positive + False negative}} \times 100$ **Answer :** (a)
(b) $\frac{\text{True negative}}{\text{True negative + False positive}} \times 100$
(c) $\frac{\text{True positive}}{\text{True positive + False positive}} \times 100$
(d) None of the above.

Q. 618. Level of which constituent of blood may vary with albumin concentration?

(a) Plasma glucose **Answer :** (c)
(b) Serum uric acid
(c) Serum calcium
(d) Serum creatinine.

Q. 619. Which one is hypoxanthine?

(a) 6-oxy purine **Answer :** (a)
(b) 2-oxy purine
(c) 4-amino purine
(d) 6-amino purine.

13 QUESTIONS ON CLINICAL BIOCHEMISTRY

Q. 620. Which is not true for hypernatremia?

(a) It is the commonest cause of increased tonicity of body fluids

(b) It is almost always due to water deficit

(c) ICF volume is decreased

(d) Hypernatremia with increased body sodium is relatively very common cause of hypernatremia.

Answer : (d)

Q. 621. Which one is not a cause of hypokalemia?

(a) Villous adenoma of colon

(b) Renal tubular acidosis

(c) Bartter's syndrome

(d) Chronic interstitial nephritis.

Answer : (d)

Q. 622. Find out the incorrect statement?

(a) Acidotic states are generally associated with hypokalemia

(b) Hypertonic states may cause hypokalemia

(c) Crush injury may cause hyperkalemia

(d) Lack of insulin can cause hypokalemia.

Answer : (a)

Q. 623. Which one of the value is abnormal for an healthy adult?

(a) Serum sodium—142 meq/L

(b) Serum potassium—3·9 meq/L

(c) Plasma bicarbonate—20 meq/L

(d) Serum urea—28 mg/dl.

Answer : (c)

Q. 624. Which one is not a cause of acidosis?

(a) Chronic renal failure

(b) Primary aldosteronism

(c) Methanol poisoning

(d) Pancreatic fistula.

Answer : (b)

Q. 625. Which is the normal value of creatinine clearance of an adult in the age group of 20—40?

(a) 80 ml/min (b) 110 ml/min

(c) 60 ml/min (d) 140 ml/min.

Answer : (b)

Q. 626. Which is not true for plasma creatinine?

(a) Plasma creatinine normally remains fairly constant through out adult life

(b) Plasma creatinine correlates well with GFR

Answer : (c)

(c) Creatinine clearance provides better precision than estimation of plasma creatinine to assess progress of renal disease
(d) Cimetidine may cause rise in plasma creatinine level.

Q. 627. Which type of protein may be detected in urine as a result of overflow proteinuria in myelomonocytic leukemia?

(a) Hemoglobin
(b) Myoglobin
(c) Lysozyme
(d) Bence Jones protein.

Answer : (c)

Q. 628. Below what level of GFR a patient of chronic renal failure cannot tolerate normal dietary sodium?

(a) GFR less than 30 ml/min
(b) GFR less than 40 ml/min
(c) GFR less than 20 ml/min
(d) GFR less than 10 ml/min.

Answer : (d)

Q. 629. At what level of GFR in chronic renal failure potassium retention leading to hyperkalemia is likely to occur?

(a) GFR 5 ml/min
(b) GFR 15 ml/min
(c) GFR 25 ml/min
(d) GFR 35 ml/min.

Answer : (a)

Q. 630. Which is the most abundant mineral in the body?

(a) Calcium
(b) Magnesium
(c) Manganese
(d) Iron.

Answer : (a)

Q. 631. Which is not true for calcitonin?

(a) It can decrease plasma calcium ion concentraction
(b) It can increase osteoclast activity
(c) It decreases renal reabsorption of calcium & phosphate
(d) Its effects are transient.

Answer : (b)

Q. 632. Which statement is incorrect with regard to plasma phosphate?

(a) Plasma phosphate below 1·2 mg/dl is associated with widespread cellular dysfunction & even death
(b) In metabolic & respiratory alkalosis phosphate moves into the cells
(c) Hypophosphatemia of diabetes ketoacidosis may worsen when insulin is administered
(d) Antacids may bind to phosphate.

Answer : (b)

Q. 633. Which is not a cause of hypophosphatemia?

(a) Chronic renal failure
(b) Vitamin D toxicity
(c) Primary PTH excess
(d) Phosphate enemas.

Answer : (c)

Q. 634. Which one is incorrect for rickets & osteomalacia caused due to deficient intake of vit. D ?

(a) Serum calcium is decreased or normal
(b) Serum phosphate may be decreased or normal

Answer : (c)

(c) Serum alkaline phosphatase is decreased
(d) Parathormone level may be increased or normal.

Q. 635. Which is not true for osteoporosis?

(a) Serum calcium is decreased
(b) Serum phosphate is normal
(c) Serum parathormone is normal
(d) Serum alkaline phosphatase is normal.

Answer : (a)

Q. 636. Which type of electrophoretic mobility is exhibited by haptoglobin?

(a) As movement of pre albumin
(b) As movement of α_1 globulin
(c) As the movement of α_2 globulin
(d) As the movement of β globulin.

Answer : (c)

Q. 637. Which type of electrophoretic mobility is exhibited by transferrin?

(a) α_1 globulin
(b) α_2 globulin
(c) β globulin
(d) γ globulin.

Answer : (c)

Q. 638. C—reactive protein belongs to which group of protein?

(a) α-globulin
(b) β-globulin
(c) γ-globulin
(d) Albuminoid.

Answer : (b)

Q. 639. Bence Jones protein found in urine of multiple myeloma consists of which one?

(a) Monomers of light chains of immunoglobulin
(b) Dimers of light chains of immunoglobulin
(c) Fragments of heavy chain of one class of immunoglobulin
(d) Fragments of both heavy & light chain of immunoglobulin.

Answer : (b)

Q. 640. In which one of the following diseases commonly heavy chain fragments of Fc portion of immunoglobulins are found in urine?

(a) Multiple myeloma
(b) Waldenstrom's macroglobulinemia
(c) Franklin's disease
(d) Benign paraproteinemia.

Answer : (c)

Q. 641. Which of the following tumour markers can be used to monitor response to treatment & detection of early recurrence of small cell lung cancer?

(a) Carcino embryonic antigen
(b) Alpha fetoprotein
(c) Neurone specific enolase
(d) Calcitonin.

Answer : (c)

Q. 642. Which is not true for creatine kinase (CK) enzyme?

(a) Most of creatine kinase (CK) belongs to CK—MM isoenzyme
(b) CK—BB rarely appears in plasma
(c) 70-80% of CK belongs to CK—MB class in cardiac muscles
(d) Plasma normally has 95% of CK as CK—MM.

Answer : (c)

Q. 643. In liver disease, plasma GGT activity generally increases in parallel with which one of the enzymes?

(a) AST
(b) ALT
(c) Alkaline phosphatase
(d) Choline esterase.

Answer : (c)

Q. 644. Estimation of which one of the enzymes is of value in organophosphorus insecticide poisoning?

(a) α-amylase
(b) Choline esterase
(c) GGT
(d) SGPT.

Answer : (b)

Q. 645. Which is not true for bilirubin?

(a) About 80% of bilirubin formed each day arises from heme of RBCs
(b) Myoglobin, cytochromes & peroxidase also contribute to total bilirubin formation
(c) Prophyrin ring remains intact in bilirubin
(d) Iron of heme is removed during bilirubin formation.

Answer : (c)

Q. 646. Which statement is incorrect?

(a) In most types of cirrhosis, plasma IgA is increased
(b) In biliary cirrhosis plasma IgM is mostly increased
(c) In chronic active hepatitis plasma IgG tends to be increased
(d) Normally, nearly 25% of bilirubin in plasma is unconjugated.

Answer : (d)

Q. 647. Which is not true for amylase?

(a) Plasma amylase is elevated in acute pancreatitis
(b) Plasma amylase may be elevated in mesenteric infarction
(c) Injection of morphine reduces plasma amylase activity
(d) In acute pancreatitis, plasma amylase level returns to normal in 3 to 5 days.

Answer : (c)

Q. 648. Which is the incorrect statement?

(a) Plasma gastrin is elevated in gastrinoma
(b) Increased plasma gastrin may be seen in patients with hypercalcemias
(c) Plasma gastrin is increased in duodenal ulcer
(d) G-cell hyperplasia causes increase in plasma gastrin.

Answer : (c)

Q. 649. Which unabsorbed amino acid, when metabolised in liver produces indican in urine?

(a) Tryptophan
(b) Methionine
(c) Cysteine
(d) Tyrosine.

Answer : (a)

Q. 650. Under normal conditions how much of dietary tryptophan is metabolised to 5 hydroxy tryptophan, 5-hydroxy tryptamine & 5 hydroxy indole acetic acid?

(a) 1% (b) 2%
(c) 5% (d) 10%.

Answer : (a)

Q. 651. In Verner Morrison syndrome hypersecretion of which G.I hormone causes watery diarrhoea & hypokalemia?

(a) Pancreatic peptide
(b) Vasoactive intestinal peptide
(c) Motilin
(d) Secretin.

Answer : (b)

Q. 652. Which biochemical abnormality is unlikely in protein energy malnutrition?

(a) Low blood sugar
(b) Rise in glucagon
(c) Rise in creatinine excretion
(d) Rise in cortisol.

Answer : (c)

Q. 653. Which is not a copper containing metalloprotein?

(a) Dopamine hydroxylase
(b) Tyrosinase
(c) Superoxide dismutase
(d) Phenylalanine hydroxylase.

Answer : (d)

Q. 654. Which one of the enzymes requires selenium?

(a) Iodothyronine deiodinase
(b) Tyrosinase
(c) Porphobilinogen synthase
(d) DNA polymerase.

Answer : (a)

Q. 655. Which of the enzymes are zinc dependent?

(a) Alkaline phosphatase
(b) Porphobilinogen synthase
(c) Alcohol dehydrogenase
(d) All of the above.

Answer : (d)

Q. 656. Which mineral deficiency is associated with Keshan's disease, a cardiomyopathy seen in some parts of China?

(a) Zinc
(b) Cobalt
(c) Selenium
(d) Iron.

Answer : (c)

Q. 657. Which mineral is present in brewer's yeast (glucose tolerance factor) which improves glucose tolerance?

(a) Nickel
(b) Chromium
(c) Cobalt
(d) Manganese.

Answer : (b)

Q. 658. Tryptophan load test can be performed to investigate which vitamin deficiency?

(a) Thiamine
(b) Pyridoxine
(c) Niacin
(d) Folic acid.

Answer : (c)

Q. 659. Besides glucose, among amino acids which one promotes insulin secretion?

(a) Cysteine
(b) Methionine
(c) Leucine
(d) Lysine.

Answer : (c)

Q. 660. What effect does vagal stimulation has on insulin release?

(a) Promotes insulin release
(b) Inhibits insulin release
(c) Has no effect on insulin release
(d) Stimulates insulin synthesis.

Answer : (a)

Q. 661. Which statement is not true?

(a) 50% of cases of gestational diabetes mellitus may develop diabetes mellitus in next 7 years
(b) Plasma glucose is lower than whole blood glucose
(c) Plasma glucose is higher than whole blood glucose
(d) In oral glucose tolerance test 82·5 gm of glucose monohydrate should be given orally after an overnight fast.

Answer : (b)

Q. 662. Which statement is incorrect?

(a) Plasma frutosamine is a measure of glycated plasma proteins
(b) Homoglobin can react with glucose, glucose 6 phosphate to produce glycated product
(c) Glycated hemoglobin with glucose is called HbA_1
(d) Glycated Hb reflects average level of blood glucose over previous 1-2 months.

Answer : (c)

Q. 663. Which one is associated with Tangier's disease?

(a) Increased rate of apo A-I catabolism
(b) HDL is increased in plasma
(c) LDL is increased in plasma
(d) Plasma cholesterol esters are reduced in lymphoreticular system.

Answer : (a)

Q. 664. Which is not true of plasma iron?

(a) Iron level in plasma shows diurnal variation with higher values in the morning
(b) Just before menstrual cycle, plasma iron increases
(c) Oral contraceptives cause increase in plasma iron
(d) Measument of plasma iron is not adequate index of iron status in the body.

Answer : (b)

Q. 665. On a routine basis which is the most useful method to measure iron status of the body?

(a) Serum iron **Answer : (d)**
(b) Plasma iron
(c) Serum transferrin
(d) Plasma ferritin.

Q. 666. In a person who develops iron deficiency which event occurs first?

(a) Plasma ferritin falls **Answer : (a)**
(b) Plasma transferrin & TIBC increase
(c) Plasma iron falls
(d) Anemia becomes evident.

Q. 667. Which one is decreased in lead poisoning?

(a) Blood lead level **Answer : (b)**
(b) Erythrocyte PBG synthetase activity
(c) Urinary ALA excretion
(d) Zn protopophyrin level in RBC.

Q. 668. How much methemoglobin should be there in blood for cyanosis to develop?

(a) 0·5 gm/dl (b) 1 gm/dl **Answer : (c)**
(c) 2 gm/dl
(d) Cyanosis does not develop in presence of methemoglobinemia.

Q. 669. Which is not true for sulphemoglobin?

(a) Sulphemoglobin cannot act as O_2 carrier **Answer : (c)**
(b) It cannot be converted back to Hb
(c) Large amount of sulphemoglobin is required to produce cyanosis
(d) H_2S produced by intestinal bacteria may help under certain circumstances to produce sulphemoglobin.

Q. 670. Allopurinol is an isomer of which compound?

(a) Xanthine (b) Hypoxanthine **Answer : (b)**
(c) Adenosine (d) Dihydroxyadenine.

Q. 671. Reverse T_3 (rT_3) is produced by which mechanism?

(a) Outer ring (5') mono-deiodination of T_4 **Answer : (b)**
(b) Inner ring (5) mono-deiodination of T_4
(c) Inter change of iodine positions of T_3 in non-thyroid tissues
(d) Coupling of mono & diodo tyrosine residues.

Q. 672. In which of these euthyroid conditions, TSH in plasma may be elevated?

(a) During first 20 weeks of pregnancy **Answer : (c)**
(b) Treatment with dopaminergic drugs
(c) Intake of amiodarone
(d) Treatment with high doses of glucocorticoids.

Q. 673. Pick out the correct statement.

(a) Free thyroid hormone level increases in pregnancy **Answer** : (c)
(b) Plasma TBG concentration decreases in pregnancy
(c) Heterophilic antibodies interfere in TSH immunoassays
(d) TSH is not a reliable indicator of thyroid state in 2nd & 3rd timester of pregnancy.

Q. 674. Which is not correct for thyrotrophin receptor antibodies (TRAbs)?

(a) TRAbs are IgG class antibodies **Answer** : (d)
(b) They are directed against TSH receptors in the thyroid gland
(c) TRAbs are positive in 90% of patients of Grave's disease
(d) TRAbs are always detectable in thyroiditis.

Q. 675. Which tumor is not associated with MEN—II a?

(a) Medullary carcinoma of thyroid **Answer** : (d)
(b) Pheochromocytoma
(c) Hyperparathyroidism
(d) Pituitary tumours.

Q. 676. Cortisol shows diurnal variation. For a person of normal night sleeping habit, when is the cortisol level highest?

(a) Start of working day (b) Post lunch hour **Answer** : (a)
(c) Evening 6 P.M. (d) On onset of sleep.

Q. 677. What is true for insulin hypoglycemia test?

(a) It identifies integrity of hypothalamic-pituitary—adrenal (HPA) axis **Answer** : (a)
(b) Should be performed in epilepsy to identify the cause
(c) Can be performed in patients of heart disease
(d) Does identify Cushing's syndrome.

Q. 678. Which combination of tests provide almost 100% specificity & sensitivity in the diagnosis of Cushing's disease?

(a) High dose dexamethasone suppression test & CRH test **Answer** : (a)
(b) Cortisol & ACTH estimation in plasma
(c) Cortisol & short synacthen test
(d) Insulin hypoglycemia test & CRH test.

Q. 679. Which one of the tests is performed to investigate adrenocortical hypofunction?

(a) Short tetracosactrin test **Answer** : (a)
(b) Low dose dexamethasone test
(c) High dose dexamethasone test
(d) CRH stimulation test.

Q. 680. Which is not true of TRH test?

(a) In primary hyperthyroidism there is no increase of TSH in response to TRH **Answer** : (c)
(b) In hypothyroidism the response of TSH to TRH stimulation is exaggerated

(c) Patients with T_3 receptor defects do not show TSH response in a TRH test
(d) In normal individuals after intravenous TRH, serum TSH increases in 20 minutes time.

Q. 681. α-Subunit of TSH is common to which one of the hormones?

(a) FSH **Answer : (a)**
(b) ACTH
(c) Growth hormone
(d) Prolactin.

Q. 682. Which is incorrect for gonadotropin releasing hormone (Gn-RH) test?

(a) Gn-RH test is used to investigate gonadotropin reserves **Answer : (b)**
(b) In clinical hypogonadism this test is always abnormal
(c) It need not be performed in primary gonadal failure with raised gonadotropin level
(d) Subnormal response in Gn-RH test is seen in Cushing's syndrome.

Q. 683. Which is not true of prolactin?

(a) Prolactin is secreted by lactotrophs **Answer : (c)**
(b) Prolactin has a diurnal rhythm of secretion
(c) Dopamine stimulates prolactin secretion
(d) Highest prolactin secretion occurs during sleep.

Q. 684. Which drug reduces prolactin secretion?

(a) Melthyl dopa **Answer : (d)**
(b) Reserpine
(c) Cimetidine
(d) Dopamine.

Q. 685. Testosterone is produced by which cell?

(a) Sertoli cells **Answer : (b)**
(b) Leydig cells
(c) Fibroblast of seminiferus tubules
(d) Tubular cells of seminiferus tubules.

Q. 686. Inhibin, a regulatory protein of testes in produced by which cell?

(a) Leydig cells **Answer : (b)**
(b) Sertoli cells
(c) Macrophages
(d) B cells of testes.

Q. 687. How much of testosterone is bound to albumin in plasma?

(a) 60% **Answer : (b)**
(b) 30%
(c) 10%
(d) Nil.

Q. 688. Peak blood levels of maternal hCG is found during which period in pregnancy?

(a) 4 weeks of pregnancy
(b) 6 weeks of pregnancy
(c) 8 weeks of pregnancy
(d) 12 weeks of pregnancy.

Answer : (d)

Q. 689. In males, which is the commonest tumor that produces hCG?

(a) Seminoma
(b) Testicular teratomas
(c) Carcinoid tumours
(d) Oat cell carcinoma.

Answer : (b)

Q. 690. Which hormone level is reduced in mother during pregnancy?

(a) Prolactin
(b) Oestrogen
(c) Testosterone
(d) Growth hormone.

Answer : (d)

Q. 691. Which is the incorrect statement?

(a) Trophoblastic tumors secrete hCG
(b) Placenta can synthesize oestriol de novo
(c) Foetal sterol such as DHAS is converted to oestriol by placenta also
(d) Oestriol production thus requires both foetal & placental components.

Answer : (b)

Q. 692. Which is true in pregnancy for mother?

(a) Plasma triglyceride level in mother falls during pregnancy
(b) Plasma fibrinogen level falls
(c) Plasma ceruloplasmin increases
(d) Plasma 1,25 dihydro cholecalciferol decreases.

Answer : (c)

Q. 693. Which protein is screened in mother to identify neural tube defect in foetus?

(a) α_1—Fetoprotein
(b) CEA
(c) Alkaline phosphatase
(d) Neural protein D.

Answer : (a)

Q. 694. Which is the commonest congenital cause of mental retardation?

(a) Phenylketonuria
(b) Down's syndrome
(c) Klinefelter's syndrome
(d) Turner's syndrome.

Answer : (b)

Q. 695. During first couple of weeks of life which is the best screening test for cystic fibrosis?

(a) Pilocarpine stimulation test
(b) Measurement of sodium in sweat
(c) Measurement of chloride in sweat
(d) Immunoreactive trypsin measurement.

Answer : (d)

Q. 696. Neuoblastomas in children accounts for one third of childhood deaths from malignant diseases. Which substance is usually constantly produced by these tumours?

(a) Dopamine (b) Acetylcholine **Answer :** (a)
(c) Serotonin (d) GABA.

Q. 697. In an elderly person if there is an isolated increase of alkaline phosphatase activity, which disease should be suspected first?

(a) Multiple myeloma **Answer :** (b)
(b) Paget's disease
(c) Carcinoma of prostate
(d) Osteosarcoma.

Q. 698. Pick out the disease with single gene defect?

(a) Coronary heart disease **Answer :** (d)
(b) Diabetes mellitus
(c) Schizophrenia
(d) Congenital adrenal hyperplasia.

Q. 699. Pick out the genetic disorder which is polygenic in inheritance?

(a) Multiple endocrine neoplasia **Answer :** (b)
(b) Manic depressive psychosis
(c) Wilson's disease
(d) Goucher's disease.

Q. 700. Which is not true of CSF glucose?

(a) Lumbar CSF glucose is lower than plasma glucose **Answer :** (d)
(b) CSF glucose may be normal in viral encephalitis
(c) In hypoglycemia, CSF glucose may become very low
(d) CSF of cerebral ventricles have same glucose as that of lumbar CSF.

Q. 701. Which is not true of CSF protein?

(a) Lumbar CSF protein is normally almost all albumin **Answer :** (c)
(b) A value of lumbar CSF protein of 90 gm/dl may be normal in neonatal period
(c) In demyelinating diseases, CSF protein decreases
(d) CSF normally contains small amounts of IgG.

Q. 702. Which one is an active treatment principle of Benzodiazepine over-dose?

(a) Flumazenil **Answer :** (a)
(b) N—acetyl cysteine
(c) Naloxone
(d) Atropine.

Q. 703. In overdose of paracetamol which substance makes active treatment principle?

(a) Flumazenil **Answer :** (b)
(b) N—acetyl cysteine
(c) Naloxone
(d) Hydroxycobalamine.

Q. 704. Toxicity of which one of these compounds can produces hypocalcemia?

(a) Ethanol **Answer** : (c)
(b) Methanol
(c) Ethylene glycol
(d) None of the above.

Q. 705. Dialysis dementia is caused by which metal toxicity?

(a) Magnesium **Answer** : (b)
(b) Aluminium
(c) Lead
(d) Manganese.

Q. 706. In a gaussian distribution what percentage of results lie in the range mean ± 1 SD?

(a) 25% (b) 52% **Answer** : (a)
(c) 67% (d) 83%.

Q. 707. Among the following secretion which has maximum potassium ions?

(a) Saliva **Answer** : (a)
(b) Gastric juice
(c) Pancreatic juice
(d) Intestinal juice.

Q. 708. Which one among the following secretions has maximum hydrogen ions?

(a) Saliva **Answer** : (b)
(b) Gastric juice
(c) Pancreatic juice
(d) Intestinal juice.

Q. 709. Which statement is incorrect?

(a) Plasma urea does not contribute much to plasma tonicity **Answer** : (c)
(b) Increase in tonicity of ECF causes reduction in ICF volume
(c) Plasma alcohol contributes to tonicity & plasma osmolality almost equally
(d) Alterations in plasma sodium causes changes in ECF & ICF volume.

Q. 710. Which is incorrect for oedema?

(a) Hyproproteinemia causes oedema **Answer** : (b)
(b) Primary aldosteronism causes oedema
(c) Increased venous pressure due to cardiac failure causes oedema
(d) Low renal perfusion pressure may cause secondary hyperaldosteronism & oedema.

Q. 711. Which one of the conditions can cause pseudo-hyponatremia?

(a) Mineralocorticoid deficiency **Answer** : (b)
(b) Severe hyperlipidemia
(c) Renal tubular acidosis
(d) Chronic opium intake.

Q. 712. Which is not usually a cause of hypernatremia?

(a) Conn's syndrome **Answer : (d)**
(b) Diarrhoea
(c) Diabetes insipidus
(d) Severe burns.

Q. 713. Which one of these factors does not tend to cause potassium ion to move into the cells?

(a) Hypertonicity **Answer : (a)**
(b) Alkalosis
(c) Insulin therapy
(d) None of the above.

Q. 714. What is the site of maximum reabsorption of potassium in the kidney?

(a) Proximal tubule **Answer : (a)**
(b) Loop of Henle
(c) Distal convoluted tubule
(d) Collecting duct.

Q. 715. Regarding potassium ions which statement is correct?

(a) Acute acidosis causes potassium ion excretion **Answer : (c)**
(b) Acute alkalosis causes potassium ion retention
(c) Both chronic acidosis & alkalosis cause increase in potassium excretion
(d) In Bartter's syndrome there is persistent hyperkalemia.

Q. 716. Which isotope of potassium is natural in occurrence?

(a) ^{38}K (b) ^{39}K **Answer : (c)**
(c) ^{40}K (d) ^{42}K.

Q. 717. Which cell is the main site of hydrogen ion formation in blood?

(a) Neutrophil **Answer : (c)**
(b) Lymphocyte
(c) RBC
(d) Macrophages.

Q. 718. Match the following during early phase of acidosis or alkalosis.

(1) Respiratory acidosis **Answer :** 1. (d); 2. (c); 3. (a); 4. (b)
(2) Respiratory alkalosis
(3) Metabolic acidosis
(4) Metabolic alkalosis.
(a) Plasma (H^+) rise and (HCO^-_3) falls
(b) Plasma (H^+) fall and (HCO^-_3) rises
(c) Plasma (H^+) and (HCO^-_3) both fall
(d) Plasma (H^+) and (HCO^-_3) both increase.

Q. 719. Which is the commonest cause of hyperchloremic acidosis?

(a) Excessive vomiting **Answer : (c)**
(b) Diabetic ketoacidosis
(c) Chronic renal failure
(d) Methanol toxicity.

Q. 720. If plasma is electrophoresed where would fibrinogen band appear?

(a) Just near albumin **Answer** : (d)
(b) Between α_1 & α_2 globulin
(c) Between α_1 & β globulin
(d) Between β & γ globulin.

Q. 721. According to electrophoretic mobility thyroid binding globulin belongs to which class?

(a) α_1 globulins **Answer** : (b)
(b) α_2 globulins
(c) β globulins
(d) γ globulins.

Q. 722. Which statement is incorrect for lactate dehydrogenase (LD) enzyme?

(a) Fast moving (in electrophoresis) LD isoenzymes are found in heart **Answer** : (d)
(b) Slow moving (in electrophoresis) LD isoenzymes are found in skeletal muscle
(c) Testes also has a minor LD
(d) All component of LD isoenzymes predominate in lungs.

Q. 723. In which one of the conditions, there may be low level of serum alkaline phosphatase?

(a) Hyperparathyrodism **Answer** : (c)
(b) Paget's disease with osteoblastic activity
(c) Hereditary vit D resistant ricket
(d) Osteomalacia.

Q. 724. Which is not true for bilirubin?

(a) In bile, 75% of bilirubin exist as monoglucuronides **Answer** : (a)
(b) Bilirubin glucuronides are degraded by bacterial action in the colon
(c) Practically most bilirubin of the plasma in normal condition is unconjugated
(d) In Gilbert's disease unconjugated hyperbilirubinemia is seen.

Q. 725. In primary biliary cirrhosis which immunoglobulin rises greatly?

(a) IgG (b) IgM **Answer** : (b)
(c) IgA (d) IgD.

Q. 726. Which salt is commonly used for urine acidification test?

(a) NH_4Cl (b) NH_4HCO_3 **Answer** : (a)
(c) $(NH_4)_2SO_4$ (d) NaH_2PO_4.

Q. 727. In Hartnup's disease intestinal epithelial transport of which amino-acid is affected most?

(a) Dicarboxylic amino acids **Answer** : (c)
(b) Dibasic amino acids
(c) Neutral amino acids
(d) Imino acids.

Q. 728. Which is true of proteinuria?

(a) Tubular proteinuria is the commonest group of persistant proteinuria **Answer : (b)**
(b) Glomerular proteinuria is the commonest group of causes of persistant proteinuria
(c) Degree of proteinuria provides an index of severity of renal disease
(d) Orthostatic proteinuria is characterised by protein in early morning sample.

Q. 729. Which is the commonest single cause of renal stones?

(a) Idiopathic hypercalciuria **Answer : (a)**
(b) Vitamin D overdose
(c) Primary hyperparathyroidism
(d) Renal tubular acidosis.

Q. 730. Which is the site at which triglycerides are transferred from mixed micelles to intestinal mucosal cells?

(a) Jejunum **Answer : (a)**
(b) Duodenum
(c) Terminal ileum
(d) Junctional region of ileum & colon.

Q. 731. Which is not true for gastrin secretion?

(a) Plasma gastrin is increased in hypochlorhydria due to pernicious anemia **Answer : (b)**
(b) Plasma gastrin is increased in duodenal ulcer
(c) Plasma gastrin is reduced in fasting state
(d) Plasma gastrin reflects the rate of gastrin secretion by pyloric antrum.

Q. 732. What is true for lactoferrin?

(a) It is secreted in milk **Answer : (c)**
(b) It is secreted in succus entericus
(c) It is secreted by pancreas
(d) It is a lactose & iron preparation.

Q. 733. In disaccharide tolerance test, after a test dose of 50gm of disacharide (either sucrose or lactose), normally plasma glucose should increase by what amount?

(a) 10 mg % **Answer : (c)**
(b) 15 mg %
(c) 20 mg %
(d) 05 mg %.

Q. 734. For a normal healthy adult taking 50 to 100 gm fat/day, what should be the fecal fat content?

(a) Upto 5 g/24 hrs. **Answer : (a)**
(b) Upto 10 g/24 hrs.
(c) Upto 12 gm/24 hrs.
(d) Fat is not excreted in stool.

Q. 735. Urinary indican test may be performed to identify which one of these?

(a) Renal tubular acidosis **Answer : (c)**
(b) Correlete GFR to tubular absorption
(c) Bacterial colonisation of small intestine
(d) Test of carbohydrate intolerance.

Q. 736. Malabsorption of fat soluble vitamins is mostly manifested by which one?

(a) Vitamin A deficiency **Answer : (b)**
(b) Vitamin D deficiency
(c) Vitamin E deficiency
(d) Vitamin K deficiency.

Q. 737. Which is not true of insulin secretion?

(a) Hyperglycemia is the most important controlling factor for insulin release **Answer : (b)**
(b) Vagal stimulus reduces insulin release
(c) Somatostatin inhibit insulin release
(d) Some amino acids also stimulate insulin release.

Q. 738. What is the normal tissue lactate : pyruvate ratio?

(a) 1 : 1 **Answer : (d)**
(b) 2 : 1
(c) 5 : 1
(d) 10 : 1.

Q. 739. Which statement is true?

(a) When glucose is oxidised aerobically to CO_2 & H_2O, there no net over production of H^+ **Answer : (a)**
(b) Lacate formation occurs when there is adequate oxygenation
(c) Lactate is produced from fat by β-oxidation process
(d) Addition of lactate anions to blood causes severe acidosis.

Q. 740. Which one of these can cause reactive hypoglycemia?

(a) Starvation **Answer : (d)**
(b) Hypothyroidism
(c) Acute pancreatitis
(d) Hereditary fructose intolerance.

Q. 741. Which is not present in glycogen storage disease?

(a) Lactic acidosis **Answer : (c)**
(b) Hyperuricemia
(c) Hyperglycemia
(d) Deposit of glycogen in liver.

Q. 742. Which one of the lipoproteins have highest phospholipid content?

(a) VLDL **Answer : (d)**
(b) IDL
(c) LDL
(d) HDL.

Q. 743. Familial hypercholesterolemia belongs to which type of hyperlipoproteinemia as per WHO classification?

(a) Type I (b) Type II a **Answer : (b)**
(c) Type III (d) Type IV & V.

Q. 744. According to WHO classification, remnant hyperlipoproteinemia belongs to which class of hyperlipoproteinemia?

(a) Type I (b) Type II **Answer : (c)**
(c) Type III (d) Type V.

Q. 745. In WHO type I hyperlipoproteinemia which enzyme may be deficient?

(a) LCAT **Answer : (c)**
(b) ACAT
(c) Lipoprotein lipase
(d) Cholesteryl esterase.

Q. 746. In osteoporosis which is unlikely to occur?

(a) Serum calcium is normal, usually **Answer : (c)**
(b) Serum phosphate is usually normal
(c) Serum alkaline phosphatase is raised
(d) Ionised calcium is normal usually.

Q. 747. In lead poisoning which enzyme activity is reduced?

(a) ALA synthase **Answer : (b)**
(b) ALA dehydratase
(c) Porphobilinogen deaminase
(d) Uroporphyrinogen III cosynthase.

Q. 748. Which is the commonest type of porphyria?

(a) Porphyria cutanea tarda **Answer : (a)**
(b) Variegate porphyria
(c) Protoporphyria
(d) Congenital erythropoetic porphyria.

Q. 749. Which statement regarding zinc is not true?

(a) Zinc is required by enzymes of nucleic acid synthesis **Answer : (c)**
(b) In plasma, zinc may be bound to albumin, transferrin
(c) Plasma zinc rises by 20%—30% after meals
(d) Body does not store appreciable amounts of zinc.

Q. 750. Deficiency of which element typically occurs in acrodermatitis enteropathica?

(a) Magnesium **Answer : (d)**
(b) Manganese
(c) Molybdenum
(d) Zinc.

Q. 751. Which one of these conditions can cause low total T_4 level in blood?

(a) Pregnancy **Answer : (c)**
(b) Oestrogen therapy
(c) Lithium therapy
(d) Hereditary TBG excess.

Q. 752. Tetracosactrin is a synthetic polypeptide with a structure identical to the N—terminal 24 amino acids of which hormone?

(a) Growth hormone (b) TSH
(c) ACTH (d) Prolactin.

Answer : (c)

Q. 753. What is the substrate of angiotensin converting enzyme (ACE)?

(a) Renin
(b) Angiotensin—I
(c) Angiotensin—II
(d) Angiotensin—III.

Answer : (b)

Q. 754. Which statement is incorrect?

(a) Rate limiting factor in the formation of angiotensin I is the renin activity in plasma
(b) Plasma renin activity (PRA) may be measured to distinguish between primary & secondary hyperaldosteronism
(c) Half life of renin in plasma is 20 minutes
(d) If PRA is high, patient has primary hyperaldosteronism.

Answer : (d)

Q. 755. Which is the commonest defect to produce congenital adrenal hyperplasia (CAH)?

(a) 21—hydroxylase
(b) 11-β hydroxylase
(c) 18 hydroxylase
(d) 3-β hydroxysteroid dehydrogenase.

Answer : (a)

Q. 756. Which enzyme converts testosterone to dehydro-testosterone (DHT)?

(a) 1-α hydroxylase
(b) 3-α hydroxylase
(c) 5-α reductase
(d) 3-α reductase.

Answer : (c)

Q. 757. Which of the statement is not correct?

(a) In hypergonadotrophic hypogonadism plasma testosterone is reduced
(b) In hypogonadotrophic hypogonadism plasma FSH, LH & testosterone may be reduced
(c) In testicular feminisation syndrome plasma testosterone is low
(d) Klinefelter's syndrome may produce hypergonadotrophic hypogonadism.

Answer : (c)

14 EXTRA QUESTIONS IN CLINICAL CHEMISTRY

Q. 758. Who invented PCR technology?

(a) Kohler & Millstein

(b) Kary Mullis

(c) Craig Venter

(d) Watson & Crick.

Answer : (b)

Q. 759. Which one of these is the cyclised form of glutamic acid?

(a) GABA

(b) β-Amino isobutyric acid

(c) Pyrrolidine carboxylic acid

(d) Taurine.

Answer : (c)

Q. 760. Which is the most common inborn error of amino acid transport leading to excetion of amino acid in urine?

(a) Classical cystinuria

(b) Melanuria

(c) Phenylketonuria

(d) Histidinuria.

Answer : (a)

Q. 761. Urocanic acid is produced by metabolism of which amino acid?

(a) Cysteine (b) Histidine

(c) Valine (d) Methionine.

Answer : (b)

Q. 762. Which is the most widely used reagent for both qualitative & quantitative assessment of amino acids?

(a) Ninhydrin

(b) Biuret

(c) Edman's reagent

(d) Fuelgen reagent.

Answer : (a)

Q. 763. What is the normal plasma half life of albumin?

(a) 5 to 10 days (b) 11 to 14 days

(c) 15 to 19 days (d) 20 to 30 days.

Answer : (c)

Q. 764. Which is not true for albumin?

(a) Its level may decrease in acute inflammation during acute phase response

(b) TNFα stimulate albumin synthesis

(c) Increase in colloidal osmotic pressure (COP) decrease albumin synthesis

(d) Alcohol may inhibit albumin synthesis in liver.

Answer : (b)

Q. 765. Which one is a negative acute phase reactant protein?

(a) Haptoglobin
(b) Transferrin
(c) C—reactive protein
(d) α_1-acid glycoprotein.

Answer : (b)

Q. 766. Which subclass of IgG is the principal of IgG to cross placenta & protect neonates for the first 3 month of postnatal life?

(a) IgG_1
(b) IgG_2
(c) IgG_3
(d) IgG_4.

Answer : (a)

Q. 767. Which type of cell helps transformation of B-lymphocytes to plasma cells?

(a) CD_{-8} lymphocytes
(b) CD_{-4} lymphocytes-subtype Th_2
(c) CD_{-4} lymphocytes-subtype Th_1
(d) Tγδ lymphocyte.

Answer : (b)

Q. 768. Which statement is not true?

(a) Th O cells represent a T-lymphocyte after antigenic stimulation
(b) Th O cells can become Th_1 cells under the influence of 1L-12 & IFN-r
(c) Th O can become Th_2 cells under the influence of IL-4
(d) Naive lymphocytes produce most types of IL_s.

Answer : (d)

Q. 769. Which one of the following has been allowed for use in hairy cell leukemia?

(a) TGF-β (b) IFNα
(c) PDGF (d) IL-6.

Answer : (b)

Q. 770. Match the columns?

(a) Interleukin—1
(b) Interleukin—2
(c) Interleukin—3
(d) Interleukin—4.

(i) B cell differentiating factor
(ii) CFU stimulating activity
(iii) B—cell activating factor
(iv) T—cell growth factor.

Answer : (d) (i), (c) (ii), (b) (iv), (a) (iii)

Q. 771. Which is not true for creatine kinase?

(a) In cerebral ischemia CK-3 isoenzyme may be raised
(b) Hypothyroidism commonly presents with diminished CK levels
(c) In normal child birth, maternal total CK rises
(d) Cardiac catheterisation can cause rise in CK value.

Answer : (b)

Q. 772. Which is not an oncofetal antigen?

(a) α-fetoprotein
(b) CEA
(c) Squamous cell antigen
(d) PSA.

Answer : (d)

Q. 773. To which plasma protein is α-fetoprotein closely related in structure?

(a) Albumin
(b) Globulin
(c) Fibrinogen
(d) Haptoglobin.

Answer : (a)

Q. 774. Which statement is not true for Carcino Embryonic Antigen (CEA)?

(a) CEA is elevated in colorectal cancer, lung cancer, breast cancer
(b) CEA may potentiate invasion & metastasis
(c) CEA is not elevated in benign diseases
(d) CEA testing should not be used for screening of cancers.

Answer : (c)

Q. 775. Pickout the incorrect statement?

(a) CA 125 is used as a marker of ovarian & endometrial cancer
(b) Oestrogen & progesterone receptors are used in ovarian & endometrial cancers as indicator for hormone therapy
(c) DU—PAN 2 is used as a marker of pancreatic cancer
(d) CA 15-3 is used as marker of breast cancer.

Answer : (b)

Q. 776. The enzyme glucose oxidase reacts with which form of glucose?

(a) α-D–Glucose
(b) β-D–Glucose
(c) α-L–Glucose
(d) β-L–Glucose.

Answer : (b)

Q. 777. Which one of these can inhibit insulin secretion?

(a) Glucose (b) Gastrin
(c) Nicotinic acid (d) Secretin.

Answer : (c)

Q. 778. Which Glucose transporter (GLUT) gene is a psuedogene and is non functional?

(a) GLUT 2 (b) GLUT 5
(c) GLUT 6 (d) GLUT 7.

Answer : (c)

Q. 779. Which of these glucose transporters in microsomal in nature i.e., release glucose from endoplasmic reticulum?

(a) GLUT 1
(b) GLUT 3
(c) GLUT 5
(d) GLUT 7.

Answer : (d)

Q. 780. Which is not true for intravenous glucose tolerance test?

(a) It is conducted in persons with poor absorption of oral glucose

(b) Done in persons with gastric resection

(c) Glucose is given i.v over a period of 15 minutes at a dose of 0·5 g/kg as solution of 25g/dl

(d) Glucose is given i.v as a 25 g/dl solution in a dose of 0·5 g/kg in a time of 3 minutes.

Answer : (c)

Q. 781. The rate of decrease of serum glucose in one hour in coagulated blood at room temperature is nearly how much?

(a) 2% to 3% (b) 3% to 5%

(c) 5% to 7% (d) 10% to 15%.

Answer : (c)

Q. 782. Which is not correct for gestational diabetes mellitus. (GDM)?

(a) Mother who have GDM have greater chance of development of diabetes mellitus later in life

(b) GDM can produce hypocalcemia, hypoglycemia & macrosomia in neonates

(c) Normal pregnancy is associated with decreased insulin resistance

(d) Screening for GDM should be done between 24 & 28 weeks of gestation in all pregnant women not identified to have glucose intolerance.

Answer : (c)

Q. 783. What is the relation between fasting whole blood glucose concentration to that of plasma in a person with a normal hematocrit?

(a) Whole blood sugar concentration is 12 to 15% more than that of plasma glucose concentration

(b) Plasma glucose concentration is 12 to 15% more than that of whole blood sugar concentration

(c) There is no difference of glucose concentration between plasma & whole blood

(d) None of the statements are correct.

Answer : (b)

Q. 784. Which statement is not correct for lactic acid?

(a) Approximately 65% of total basal lactate production is used by liver

(b) Glucose to lactate conversion & vice versa is called Cori's cycle

(c) A lactate : pyruvate ratio less than 25 suggests hypoxia

(d) A high lactate : pyruvate ratio is a sensitive test of mitochondrial muscle toxicity in zidovudine therapy.

Answer : (c)

Q. 785. Pick out the incorrect statement of lactic acidosis.

(a) Type A lactic acidosis is seen in decreased tissue oxygenation

(b) Type B lactic acidosis is metabolic in origin

(c) Hypotension increases mortality in lactic acidosis

(d) Type B lactic acidosis is much more common.

Answer : (d)

Q. 786. HbA1C is formed due to which process?

(a) Condensation of glucose with N-terminal valine residue of each β-chain of HbA

(b) Condensation of glucose with N-terminal lysine residue of each β-chain of HbA

(c) Condensation of glucose with N-terminal valine of α-chains of HbA

(d) Condensation of glucose with N-terminal lysine residue of each α-chain of HbA.

Answer : (a)

Q. 787. Pick out the correct statement?

(a) Primary clinical application of insulin measurement is evaluation of patients with fasting hypoglycemia

(b) C-peptide estimation is required for diagnosis of diabetes mellitus

(c) Insulin estimation is an essential tool for diagnosis of hyperglycemic state

(d) Insulin assesment is most helpful in familial hyperpro-insulinemia.

Answer : (a)

Q. 788. Which is not correct for fructosamine?

(a) There is no role of fructosamine assay for diagnosis of diabetes mellitius

(b) Frutosamine is the generic name of plasma protein ketoamines

(c) Glucose interacts with amino group of lysing residues to produce fructosamine

(d) Ascorbic acid at concentration of more than 5 mg/dl. in plasma may cause positive interference in fructosamine assay.

Answer : (d)

Q. 789. Microalbuminuria is said to be present if 24 hrs urine excretion of albumin exceeds how much?

(a) 5 mg

(b) 10 mg

(c) 20 mg

(d) 30 mg.

Answer : (d)

Q. 790. Which one is not an isoprene polymer?

(a) Vitamin A

(b) Vitamin E

(c) Vitamin K

(d) Vitamin D.

Answer : (d)

Q. 791. If dietary cholesterol is 3g/day in an healthy adult,.how much can maximum be absorbed?

(a) 300 mg

(b) 500 mg

(c) 1,000 mg

(d) 2,000 mg.

Answer : (c)

Q. 792. Which apolipoprotein is vital for chylomicron formation?

(a) apo B—48
(b) apo B—100
(c) apo E—I
(d) apo A—I.

Answer : (a)

Q. 793. Which one inhibits conversion of desmosterol to cholesterol (the final step)?

(a) Lanosterol
(b) Nicotinic acid
(c) Triparanol
(d) Simvastatin.

Answer : (c)

Q. 794. Which one is an eicosatetraenoic acid?

(a) Linoleic acid
(b) α-linolenic acid
(c) γ-linolenic acid
(d) Arachidonic acid.

Answer : (d)

Q. 795. Which of these unsaturated fatty acid produce prostaglandin—I (PG-1) series compounds?

(a) Eicosatrienoic acid
(b) Eicosatetraenoic acid
(c) Eicosapentaenoic acid
(d) Docasa hexaenoic acid.

Answer : (a)

Q. 796. Which is the rate limiting step in prostaglandin synthesis?

(a) Step of prostaglandin reductase
(b) Step of prostaglandin dehydrogenase
(c) Cycloxygenase step
(d) Release C_{20} fatty acid from phospholipids.

Answer : (d)

Q. 797. In fasting state which lipoprotein carries most triglycerides?

(a) HDL
(b) LDL
(c) IDL
(d) VLDL.

Answer : (d)

Q. 798. Which protein constitutes most of the apolipoprotein in HDL?

(a) Apo–A–I
(b) Apo–A–II
(c) Apo–C
(d) Apo–E.

Answer : (a)

Q. 799. Which apolipoprotein plays an important role in VLDL secretion from liver?

(a) Apo–A–I
(b) Apo–B–100
(c) Apo–C–I
(d) Apo–E.

Answer : (b)

Q. 800. Which apolipoprotein plays a central role in VLDL remnant metabolism?

(a) Apo–A **Answer** : (d)
(b) Apo–B–100
(c) Apo–C–I
(d) Apo–E.

Q. 801. Which metal toxicity is seen in patients undergoing prolonged hemodialysis for renal failure?

(a) Copper **Answer** : (b)
(b) Aluminium
(c) Zinc
(d) Manganese.

Q. 802. Onset of parkinsonism below age 50 may occur due to which metal toxicity?

(a) Manganese **Answer** : (a)
(b) Aluminium
(c) Lead
(d) Thallium.

Q. 803. Renal disease in aerosol painters may occur due to which metal toxicity?

(a) Lead **Answer** : (c)
(b) Arsenic
(c) Cadmium
(d) Copper.

Q. 804. Study of which specimen is least useful to identify arsenic exposure?

(a) Urine (b) Serum **Answer** : (b)
(c) Hair (d) Nail.

Q. 805. Which element is known as "glucose tolerance factor"?

(a) Selenium (b) Chromium **Answer** : (b)
(c) Zinc (d) Cadmium.

Q. 806. Erythrocyte protoporphyrin is used as a marker of which metal exposure?

(a) Copper **Answer** : (c)
(b) Zinc
(c) Lead
(d) Aluminium.

Q. 807. Which one could be most toxic to humans?

(a) Aluminum hydroxide **Answer** : (d)
(b) Elemental mercury
(c) Arsenobetaine
(d) Nickel carbonyl compound.

Q. 808. Which tissue of humans is most sensitive to selenium deficiency?

(a) Brain (b) Heart **Answer** : (b)
(c) Adrenal cortex (d) Skeletal muscle.

Q. 809. Which one of these vitamins has a chromanol nucleus?

(a) Vit A
(b) Vit D
(c) Vit E
(d) Vit K.

Answer : (c)

Q. 810. Excess intake of vitamin E may suppress absorption of which vitamin?

(a) Vit A
(b) Vit K
(c) Vit C
(d) Vit B_{12}.

Answer : (b)

Q. 811. One type of seborrheic dermatitis in infants under 6 months of age can be caused by deficiency of which one?

(a) Thiamine
(b) Vitamin E
(c) Biotin
(d) Pyridoxine.

Answer : (c)

Q. 812. Which is the major organ of magnesium homeostasis in maintaining plasma magnesium?

(a) Lungs
(b) Liver
(c) Kidney
(d) Intestine.

Answer : (c)

Q. 813. Second to iron, which is the most abundant trace element in the body?

(a) Zinc
(b) Copper
(c) Selenium
(d) Molybdenum.

Answer : (a)

Q. 814. Zinc is transported in plasma mostly as which form?

(a) Free ionic form
(b) Bound to albumin
(c) Bound to IgG
(d) Bound to IgA.

Answer : (b)

Q. 815. Reduced taste acuity is seen with defeciency of which element?

(a) Copper
(b) Selenium
(c) Zinc
(d) Iron.

Answer : (c)

Q. 816. Abnormal dark adaptation may be produced due to deficiency of which one?

(a) Zinc
(b) Copper
(c) Selenium
(d) Iron.

Answer : (a)

Q. 817. Metalloenzymes like carboxypeptidase, alcohol dehydrogenase, alkaline phosphatase & RNA polymerase require which metal?

(a) Copper **Answer** : (b)
(b) Zinc
(c) Magnesium
(d) Manganese.

Q. 818. Menke's syndrome, an X-linked disease is due to deficiency of which element?

(a) Iron **Answer** : (b)
(b) Copper
(c) Zinc
(d) Selenium.

Q. 819. Which one of these tend to lower serum copper?

(a) Pregnancy **Answer** : (c)
(b) Inflammatory conditions
(c) Adrenocorticotropic drug therapy
(d) Birth control pills.

Q. 820. Which one of the enzymes is not molybdenum dependent?

(a) Xanthine oxidase **Answer** : (d)
(b) Sulphite oxidase
(c) Aldehyde oxidase
(d) Amino acid oxidase.

Q. 821. Which one of the enzymes is not selenium dependent?

(a) Glutathione peroxidase **Answer** : (d)
(b) Iodothyronine deiodinase
(c) Thioredoxin reductase
(d) Thymidylate kinase.

Q. 822. Kasin Beck disease, an endemic osteoarthritis exhibits low levels of which element?

(a) Molybdenum **Answer** : (c)
(b) Copper
(c) Selenium
(d) Zinc.

Q. 823. Stimulation of post translational modification of collagen could be effected by which of these?

(a) Boron **Answer** : (d)
(b) Vanadium
(c) Nickel
(d) Silicon.

Q. 824. Which one of the following conditions decrease affinity of hemoglobin for oxygen?

(a) Temperature more than 37°C **Answer** : (a)
(b) Blood pH more the 7·4
(c) Blood pCO_2 less then 40 mm of Hg
(d) Reduction of 2, 3 DPG in RBC.

Q. 825. Which one of these does not cause hypokalemia?

(a) Insulin therapy in diabetic hyperglycemia
(b) Alkalosis
(c) Hypothermia
(d) Thrombocytosis.

Answer : (d)

Q. 826. Which one is not a cause of hypochloremia?

(a) Bromide intoxication
(b) Diabetic ketoacidosis
(c) SIADH
(d) Renal tubular acidosis.

Answer : (d)

Q. 827. Which one of these metabolic acidotic conditions presents with normal anion gap?

(a) Methanol toxicity
(b) Uremia of renal failure
(c) Lactic acidosis
(d) Renal tubular acidosis.

Answer : (d)

Q. 828. Which one of these metabolic acidotic conditions is associated with high anion gap?

(a) Severe diarrhoea
(b) Intestinal fistula
(c) Renal tubular acidosis
(d) Starvation.

Answer : (d)

Q. 829. Intake of which compound may lead to precipitation of calcium oxalate & hippurate in urinary tract?

(a) Salicylate
(b) Ethylene glycol
(c) Paraldehyde
(d) Methanol.

Answer : (b)

Q. 830. Which value of serum to ascitic fluid albumin gradient is diagnostic of ascites due to portal hypertension?

(a) > 0·3 g/dl (b) > 0·5 g/dl
(c) < 1·1 g/dl (d) > 1·1 g/dl.

Answer : (d)

Q. 831. Which of the statements is untrue?

(a) In man, daily production of bilirubin from all sources is 250-300 mg
(b) Heme from destruction of senescent RBC accounts for 95% of the total bilirubin production
(c) Biliverdin is reduced to produce bilirubin
(d) Heme when breaks down to form biliverdin by heme oxidase produce carbon monoxide.

Answer : (b)

Q. 832. Aminopyrine clearance test & caffeine clearance test is used to assess the function of which organ?

(a) Brain (b) Liver
(c) Kidney (d) Spleen.

Answer : (b)

Q. 833. About how much of albumin is synthesized per day in an healthy adult?

(a) 2 g (b) 5 g **Answer : (c)**
(c) 10 g (d) 15 g.

Q. 834. Which is the major source of circulating ammonia?

Answer : (b)

(a) Liver
(b) G. I. tract
(c) Lungs
(d) Kidney.

Q. 835. Ammonia can enter the CNS tissues by which process?

Answer : (a)

(a) Passive diffussion
(b) Active transport
(c) Facilitated diffusion
(d) Carrier mediated process.

Q. 836. Breast milk hyperbilrubinemia is due to which one of these?

Answer : (c)

(a) Presence of bilrubin in breast milk
(b) Infected breast milk by hepatitis G. virus
(c) Beta glucuronidase in breast milk
(d) X-linked recessive disorder.

Q. 837. Which one of these viruses is stated to be an incomplete virus & requires the help of another virus for survival?

Answer : (b)

(a) Hepatitis B Virus
(b) Hepatitis D Virus
(c) Hepatitis E Virus
(d) Hepatitis G Virus.

Q. 838. Which one of these is a single stranded anti-sense RNA virus?

Answer : (c)

(a) Hepatitis A Virus
(b) Hepatitis C Virus
(c) Hepatitis D Virus
(d) Hepatitis E Virus.

Q. 839. Which one of these hepatitis viruses is most innocuous in nature to produce hepatitis?

Answer : (d)

(a) Hepatitis B Virus
(b) Hepatitis D Virus
(c) Hepatitis E Virus
(d) Hepatitis G Virus.

Q. 840. Which statement is not correct for alcoholic liver disease?

Answer : (b)

(a) For most people 80g of alcohol (20 ml of whisky) per day is the risk dose to develop liver disease
(b) Intermittent drinking is more dangerous than daily drinking
(c) 10 to 15% of chronic alcoholics may develop cirrhosis
(d) Both hepatitis B & Hepatitis C appear to accelerate the course of alcoholic liver disease.

Q. 841. Which one of these is not the cause of cirrhosis of liver?

(a) Methotraxol
(b) Alcohol
(c) Urosodeoxy cholic acid
(d) α_1—Antitrypsin deficiency.

Answer : (c)

Q. 842. How much fat by weight should be present in liver to produce fatty liver?

(a) > 1% by weight
(b) > 2% by weight
(c) > 3% by weight
(d) > 5% by weight.

Answer : (d)

Q. 843. Which one of these can act as a nucleating factor (provide nidus) for cholesterol stone formation?

(a) Mucin
(b) Apolipoprotein–A–I
(c) Apolipoprotein–A–II
(d) Lecithin.

Answer : (a)

Q. 844. Which one of these rises earliest in serum after an attack of myocardial infarction?

(a) Myoglobin
(b) CK—MB
(c) AST
(d) LDH.

Answer : (a)

Q. 845. Which one of these can increase GFR?

(a) Increased plasma oncotic pressure
(b) Increased intratubular pressure
(c) Afferent arteriolar constriction
(d) Efferent arteriolar constriction.

Answer : (d)

Q. 846. Which one of these can have a positive effect on GFR?

(a) Adenosine
(b) Endothelin
(c) Atrial natriuretic factor
(d) Nitric oxide.

Answer : (c)

Q. 847. Which of these is used as a marker of proximal tubular damage?

(a) Albumin
(b) Retinol binding protein
(c) Transferrin
(d) IgG.

Answer : (b)

Q. 848. At what value of GFR, it may generally be expected that parathermone secretion would be increased during chronic renal failure?

(a) Below 90 ml/min/1·73 m^2
(b) Below 75 ml/min/1·73 m^2
(c) Below 55 ml/min/1·73 m^2
(d) Below 45 ml/min/1·73 m^2.

Answer : (d)

Q. 849. Which one of the following is not seen in renal tubular acidosis?

(a) Normal anion gap **Answer :** (d)
(b) Urinary bicarbonate excretion inappropriate for plasma pH
(c) Urinary proton excretion inappropriate for plasma pH
(d) Hypochloremia.

Q. 850. Which one of the following drugs can produce nephrogenic diabetes insipidius?

(a) Cisplatin (b) Lithium **Answer :** (b)
(c) Allopurinol (d) ACE inhibitors.

Q. 851. Ammonium chloride loading test is performed in the diagnosis of which one?

(a) Distal renal tubular acidosis **Answer :** (a)
(b) Renal basement membrane disease
(c) Hepatorenal syndrome
(d) Nephrogenic diabetes insipidus.

Q. 852. Which type of renal stone is the commonest type of renal stone?

(a) Calcium oxalate with or without phosphate **Answer :** (a)
(b) Triple phosphate
(c) Calcium phosphate
(d) Urate.

Q. 853. Which one of these conditions is not associated with fasting hypergastrinemia?

(a) Infection with helitobactor pylori **Answer :** (b)
(b) Duodenal ulcer
(c) Older people beyond 60 years of age
(d) Chronic renal failure with secondary hyperparathyroidism.

Q. 854. Cholecystokinin secretion is completely inhibited by infusion of which one?

(a) Insulin (b) Somatostatin **Answer :** (b)
(c) Gastrin (d) Tryptophan.

Q. 855. Mucosal granular S cells which secrete secretin are located in greatest concentration in which site?

(a) Greater curvature of stomach **Answer :** (b)
(b) Duodenum
(c) Jejunum
(d) Lower intestine.

Q. 856. Verner Morrison syndrome (also called pancreatic cholera) is associated with increased plasma concentration of which one?

(a) VIP (b) Secretin **Answer :** (a)
(c) GIP (d) CCK.

Q. 857. Intrinsic factor which binds to vitamin B_{12} is secreted from which part of stomach?

(a) Antrum (b) Fundus **Answer :** (b)
(c) Lesser curvature (d) Greater curvature.

15 MORE QUESTIONS IN BIOCHEMISTRY

Q. 858. Homocysteine is produced from which amino acid?

(a) Cysteine
(b) Methionine
(c) Tryptophan
(d) Histidine.

Answer : (b)

Q. 859. In Lac operon system in E. coli lactose acts as which type of regulator for the operon?

(a) Repressor
(b) Derepressor
(c) Operator
(d) Promoter.

Answer : (b)

Q. 860. Barbiturates can act as a derepressor of which enzyme?

(a) Transaminases
(b) Tryptophan pyrrolase
(c) ALA synthase
(d) Glucokinase.

Answer : (c)

Q. 861. Which of these is not a DNA virus?

(a) Herpes virus
(b) Small pox virus
(c) Adeno virus
(d) Corona virus.

Answer : (d)

Q. 862. Acyclovir is an analogue of which one?

(a) Adenosine
(b) Guanosine
(c) Cytosine
(d) Uracil.

Answer : (b)

Q. 863. Neurofibril tangles in CNS is a typical finding in which disease?

(a) Scrapie
(b) Kuru
(c) Alzheimer's disease
(d) Fatal familial insomnia.

Answer : (c)

Q. 864. Which is one of the major susceptibility gene for Alzheimer's disease?

(a) Apo—E_1 (b) Apo—E_2
(c) Apo—E_3 (d) Apo—E_4.

Answer : (d)

Q. 865. Which one is inherited as a X-linked disease?

(a) Severe combined immunodeficiency **Answer** : (b)
(b) Duchenne muscular dystrophy
(c) Cystic fibrosis
(d) Familial hypercholesterolemia.

Q. 866. In reverse PCR method, enzyme of which bacteria is used?

(a) Thermus aquaticus **Answer** : (b)
(b) Thermus thermophilus
(c) Thermus megaphilus
(d) Thermus normophilus.

Q. 867. About what percent of population are resistant to HIV infection?

(a) 1% **Answer** : (a)
(b) 2%
(c) 0·1%
(d) 0·5%.

Q. 868. Regarding chromosomal translocation which statement is not correct?

(a) In Burkitt's lymphoma, translocation of chromosome 8 to 14 occurs **Answer** : (d)
(b) Deletion of short arm of chromosome 22 is called Philadelphia chromosome
(c) In non-Hodgkin's lymphoma, translocation of chromosome 14 to 18 may occur
(d) In chronic myeloid leukemia, translocation of 22 to 9 may occur leading to activation of C—abl gene of chromosome 9.

Q. 869. CA 125 is a tumor marker of which cancer?

(a) Breast cancer **Answer** : (b)
(b) Ovarian cancer
(c) Uterine cancer
(d) Pancreatic cancer.

Q. 870. Which radioisotope is used for teletherapy for cancer?

(a) ^{32}P **Answer** : (b)
(b) $^{137}C_S$
(c) $^{99}T_C$
(d) $^{51}C_T$.

Q. 871. Milk from which source would yield highest amount of protein per 100 ml?

(a) Human **Answer** : (b)
(b) Buffalo
(c) Cow
(d) Goat.

Q. 872. Which type of milk supplies least amount of carbohydrate per 100 ml?

(a) Human (b) Cow **Answer** : (b)
(c) Buffalo (d) Goat.

Q. 873. Casein constitutes what percent of protein in cow's milk?

(a) 60%
(b) 70%
(c) 80%
(d) 90%.

Answer : (c)

Q. 874. What should be the lecithin : sphingomyelin ratio (L/S) to identify Lungs maturity in foetus?

(a) 0·5
(b) 1
(c) 1·5
(d) 2.

Answer : (d)

Q. 875. Which one of the following has highest cholesterol content per 100 gm?

(a) Ghee
(b) Butter
(c) Milk
(d) Egg yolk.

Answer : (d)

Q. 876. What is the approximate normal BMR for an healty adult?

(a) 10 Kcal/kg/day
(b) 16 Kcal/kg/day
(c) 24 Kcal/kg/day
(d) 42 Kcal/kg/day.

Answer : (c)

Q. 877. Which one of these is not a –SH (thiol) containing compound?

(a) CoA
(b) Lipoic acid
(c) Glutathione
(d) Creatinine.

Answer : (d)

16 MORE QUESTIONS IN CLINICAL BIOCHEMISTRY

Q. 878. Which of the following refers to movement of buffer ions & solvent relative to the fixed support?

Answer : (b)

(a) Isoelectric focussing
(b) Electroendosmosis
(c) Zone electrophoresis
(d) Plasmapheresis.

Q. 879. In capillary electrophoresis which moves solvent?

Answer : (b)

(a) Cations
(b) Anions
(c) Neutral ions
(d) Solvent molecules.

Q. 880. A Gaussian distribution is usually :

Answer : (a)

(a) Bell shaped (b) Rectangular
(c) Skewed (d) Sigmoid.

Q. 881. Stokes shift is related to which one?

Answer : (c)

(a) Colorimetry
(b) Spectrophotometry
(c) Flurometry
(d) Flame photometry.

Q. 882. Which immune precipitator method may be used to quantitate a serum protein?

Answer : (a)

(a) Radial immunodiffusion
(b) Immuno fixation electrophoresis
(c) Counter immunoelectrophoresis
(d) Double diffusion.

Q. 883. Low CSF albumin may be found in which condition?

Answer : (c)

(a) Poliomyelitis
(b) Fungal meningitis
(c) Hyperthyroidism
(d) Multiple sclerosis.

Q. 884. In which condition oligoclonal banding of CSF is seen when the CSF is subjected to electrophoresis?

Answer : (b)

(a) Disk herniation
(b) Multiple sclerosis
(c) Hyperthyroidism
(d) Cerebral infarction.

Q. 885. Which enzyme level goes very high in malignant hyperthermia?

(a) CK
(b) LDH
(c) AST
(d) ALP.

Answer : (a)

Q. 886. In normal human serum, isoamylases migrate in regions corresponding to which region of protein electrophoresis?

(a) Prealbumin
(b) Albumin
(c) α—β Globulin
(d) β—γ Globulin.

Answer : (d)

Q. 887. In normal serum which isoenzyme of amylase represents nearly two third of amylase activity?

(a) Pancreatic (P) amylase
(b) P_3 isoamylase
(c) S-type isoamylase
(d) P_1 isoamylase.

Answer : (c)

Q. 888. Increased level of Glu. 6. PD enzyme is seen in :

(a) Myocardial infarction
(b) Hepatic disorders
(c) Primaquin induced hemolytic anemia
(d) Drug induced hemolytic anemias.

Answer : (a)

Q. 889. Which enzyme is most specific for β-D Glucose?

(a) Glucose oxidase
(b) Hexokinase
(c) Phosphohexose isomerase
(d) Glucose—6—phosphatase.

Answer : (a)

Q. 890. All forms of glycosylated Hemoglobin (HbA_{1a}, HbA_{1b}, HbA_{1c}) can be separated by which method?

(a) Electrophoresis
(b) Cation exchange chromatography
(c) High performance liquid chromatography
(d) Flame photometry.

Answer : (c)

Q. 891. Which one is not a feature of von Gierke's disease?

(a) Hypoglycemia
(b) Increased plasma lactate
(c) Increased urate
(d) Hypolipemia.

Answer : (d)

Q. 892. Which is not a component of intestinal lipoprotein?

(a) Apo–B–100
(b) Apo–B–48
(c) Apo–A–IV
(d) Apo–A–I.

Answer : (a)

Q. 893. Which is true for β–VLDL?

Answer : (a)

(a) Collective name of chylomicron remnants & VLDL remnants
(b) Can be cleared from plasma by binding to LDL receptor
(c) Moves to β-position in electrophoresis
(d) There is no β-VLDL terminology.

Q. 894. Beta quantification for lipids refers to quantitation of which one?

Answer : (b)

(a) HDL
(b) LDL
(c) VLDL
(d) Chylomicron.

Q. 895. Friedewald's calculation is used to indirectly quantitate which one of these?

Answer : (c)

(a) Chylomicron
(b) VLDL
(c) LDL
(d) HDL.

Q. 896. Severe hypertriglyceridemia may occur in which one?

Answer : (a)

(a) Apo C—II deficiency
(b) Apo A—I deficiency
(c) Apo B—100 deficiency
(d) Apo B—48 deficiency.

Q. 897. Normally what percentage of total NPN in plasma is contributed by plasma urea?

Answer : (c)

(a) 15%
(b) 25%
(c) 45%
(d) 65%.

Q. 898. Normally what percentage of total NPN in plasma is contributed by plasma uric acid?

Answer : (d)

(a) 5%
(b) 10%
(c) 15%
(d) 20%.

Q. 899. Normally what percentage of total NPN in plasma is contributed by plasma creatinine?

Answer : (a)

(a) 5%
(b) 10%
(c) 15%
(d) 20%.

Q. 900. A low BUN/creatinine ratio may be seen in which one?

Answer : (c)

(a) Corticosteroid therapy
(b) Gastrointestinal hemorrhage
(c) Late pregnancy
(d) High fever.

Q. 901. Which one of these is not helpful in the study of kidney disease?

(a) Urea
(b) Uric acid
(c) Ammonia
(d) Creatinine.

Answer : (c)

Q. 902. Blood ammonia level may be useful to evaluate which one of the conditions?

(a) Renal failure
(b) Acid base disorders
(c) Diabetic ketoacidosis
(d) Reye's syndrome.

Answer : (d)

Q. 903. Which is true for blood ammonia?

(a) Blood ammonia level rise rapidly after collection of blood
(b) Blood ammonia level falls rapidly after collection of blood
(c) Heparin is a preferred anticoagulant than EDTA
(d) Hemolysis does not alter ammonia content.

Answer : (a)

Q. 904. Porphyrins fluoresce which colour when excited by light near 400 nm?

(a) Yellow
(b) Red
(c) Green
(d) Blue.

Answer : (b)

Q. 905. Which is not a characteristic of porphyrinogens?

(a) They are unstable
(b) They do not fluoresce
(c) They are oxidised form of porphyrins
(d) They are difficult to be analyzed.

Answer : (c)

Q. 906. Which one is the most common type of porphyria?

(a) Porphyria cutanea tarda
(b) Variegate porphyria
(c) Hepatoerythropoetic porphyria
(d) Acute intermittent porphyria.

Answer : (a)

Q. 907. In which of these abnormal hemoglobins, the defect is not a single amino acid substitution?

(a) HbS
(b) HbC
(c) Hemoglobin D
(d) Hemoglobin lepore.

Answer : (d)

Q. 908. What percentage of hemoglobin is normally broken down by extra-vascular reticuloendothelial system?

(a) 60%
(b) 70%
(c) 80%
(d) 90%.

Answer : (d)

Q. 909. Cooley's anemia can be also called as what?

Answer : (a)

(a) β Thalassemia major
(b) β Thalassemia minor
(c) α Thalassemia major
(d) α Thalassemia minor.

Q. 910. In cellulose acetate electrophoresis in alkaline pH which type of hemoglobin shows least mobility?

Answer : (a)

(a) HbC
(b) HbS
(c) HbF
(d) HbA.

Q. 911. In cellulose acetate electrophoresis in alkaline pH which type of hemoglobin shows fastest movement?

Answer : (d)

(a) HbC
(b) HbS
(c) HbF
(d) HbA.

Q. 912. Which one of these can be termed fast hemoglobin in term of electrophoretic mobility in alkaline pH?

Answer : (c)

(a) HbS
(b) HbA
(c) HbH
(d) HbC.

Q. 913. The principle of resistance to alkali denaturation is applicable to quantitate which type of hemoglobin?

Answer : (d)

(a) HbS
(b) HbA
(c) HbH
(d) HbC.

Q. 914. A decrease in which one of these indicates intravascular hemolysis?

Answer : (d)

(a) Methemoglobin
(b) Methemalbumin
(c) Hemopexin
(d) Haptoglobin.

Q. 915. Release of more than what amount of hemoglobin can lead to appearance of free hemoglobin & methemoglobin in urine?

Answer : (d)

(a) 2g/day
(b) 3g/day
(c) 4g/day
(d) 5g/day.

Q. 916. What is the main purpose of porphyrins in human body?

Answer : (d)

(a) Carry oxygen to the tissues
(b) Synthesis of hemoglobin only
(c) Combine with free hemoglobin
(d) Synthesis of hemoglobin, myglobin & cytochromes.

Q. 917. Sodium excretion in urine beyond which amount suggests renal loss of sodium & water?

(a) 5 mmol/day
(b) 8 mmol/day
(c) 10 mmol/day
(d) 20 mmol/day.

Answer : (d)

Q. 918. Which one of the conditions can produce hypovolemia without loss of urine sodium in excess of 20 mmol/day?

(a) Diarrhoea
(b) Vomiting
(c) Excessive sweating
(d) Ketonuria.

Answer : (d)

Q. 919. What percentage of body's total potassium circulates in plasma?

(a) 2% (b) 3%
(c) 4% (d) 5%.

Answer : (a)

Q. 920. Which one of these impairs cellular entry of potassium ion?

(a) Epinephrine
(b) Digoxin overdose
(c) Hypomagnesemia
(d) Propranolol.

Answer : (d)

Q. 921. Which one of these inhibits aldosterone secretion?

(a) Heparin
(b) Spironolactone
(c) Captopril
(d) Cyclosporine.

Answer : (a)

Q. 922. Which is the second most abundant anion in the ECF?

(a) Chloride
(b) Bicarbonate
(c) Lactate
(d) Phosphate.

Answer : (b)

Q. 923. Usually what percentage of bicarbonate ion is reabsorbed by proximal tubules in the kidney?

(a) 65% (b) 75%
(c) 85% (d) 95%.

Answer : (c)

Q. 924. Which is the second most abundant intracellular cation?

(a) Magnesium
(b) Potassium
(c) Sodium
(d) Manganese.

Answer : (a)

Q. 925. In which tissue maximum amount of body magnesium is found?

(a) Bone
(b) Muscle & soft tissues
(c) Serum
(d) Liver.

Answer : (a)

Q. 926. Which is the major renal regulatory site of magnesium reabsorption?

(a) Proximal convoluted tubule
(b) Loop of Henle
(c) Distal convoluted tubule
(d) Collecting tubule.

Answer : (b)

Q. 927. Which factor can increase renal excretion of magnesium?

(a) Parathormone
(b) ANF
(c) Aldosterone
(d) Hypercalcemia.

Answer : (c)

Q. 928. Chronic hypomagnesemia may be a frequent cause of which condition?

(a) Hypocalcemia
(b) Hypokalemia
(c) Hyponatremia
(d) Hypomangnesemia.

Answer : (a)

Q. 929. What is the normal value of anion gap?

(a) 5 (b) 15
(c) 25 (d) 35.

Answer : (b)

Q. 930. Hypercalcemia is seen in all except which one condition?

(a) Hyperthyroidism
(b) Primary hyperparathyroidism
(c) Multiple myeloma
(d) Hypermagnesemia.

Answer : (d)

Q. 931. Hypocalcemia is produced in all except which one condition?

(a) Primary hypoparathyroidism
(b) Hypomagnesemia
(c) Acute pancreatitis
(d) Benign familial hypocalciuria.

Answer : (d)

Q. 932. All can cause hypokalemia except which one?

(a) Renal tubular acidosis
(b) Hyperaldosteronism
(c) Hypomagnesemia
(d) Acidosis.

Answer : (d)

Q. 933. Hyperkalemia is produced by each of the condition except?

(a) Chronic renal failure with GFR less than 20 ml/min
(b) Addison's disease
(c) Hemolysis
(d) Hypomagnesemia.

Answer : (d)

Q. 934. Hyponatremia can be produced by each of the following except which one?

(a) Aldosterone deficiency
(b) Prolonged vomiting
(c) Ketonuria
(d) Severe burn.

Answer : (d)

Q. 935. Which one of these may be classified as ultra trace element?

(a) Iron (b) Zinc **Answer : (d)**
(c) Copper (d) Iodine.

Q. 936. Superoxide dismutase enzyme contains what?

(a) Copper **Answer : (b)**
(b) Copper & zinc
(c) Zinc
(d) Selenium.

Q. 937. Supplementation of which element may improve glucose tolerance?

(a) Copper **Answer : (b)**
(b) Chromium
(c) Molybdenum
(d) Cobalt.

Q. 938. Which enzyme is not associated with manganese?

(a) Pyruvate carboxylase **Answer : (d)**
(b) Arginase
(c) Glucokinase
(d) Alkaline phosphatase.

Q. 939. Which element can cause copper deficiency?

(a) Zinc **Answer : (a)**
(b) Iron
(c) Molybdenum
(d) Selenium.

Q. 940. Deficiency of which element can cause iron deficiency?

(a) Zinc **Answer : (c)**
(b) Cobalt
(c) Copper
(d) Chromium.

Q. 941. Serum iron is most likely to be decreased in?

(a) Hemochromatosis **Answer : (d)**
(b) Viral hepatitis
(c) Sideroblastic anemia
(d) Chronic infection.

Q. 942. What is the ratio of carbonic acid to bicabonate in normal arterial blood of humans?

(a) 1 : 10 (b) 1 : 20 **Answer : (b)**
(c) 0·5 : 20 (d) 20 : 1.

Q. 943. When arterial blood from a normal person is exposed to room atmosphere, the most likely outcome would be :

(a) pCO_2 increase, pO_2 increase **Answer : (b)**
(b) pCO_2 decrease, pO_2 increase
(c) pCO_2 decrease, pO_2 decrease
(d) pCO_2 increase, pO_2 decrease.

Q. 944. Which one of the anticoagulants is preferrable for collection of arterial blood for blood gas analysis?

(a) EDTA **Answer** : (c)
(b) Sodium fluoride
(c) Lithium heparin
(d) Sodium citrate.

Q. 945. Which one is not synthesized in liver?

(a) VLDL **Answer** : (d)
(b) Phospholipid
(c) HDL
(d) LDL.

Q. 946. What is the colour of urobilinogen?

(a) Red **Answer** : (d)
(b) Pink
(c) Yellowish red
(d) Colourless.

Q. 947. 'Dane Particle' was earlier used for which virus?

(a) Hepatitis A **Answer** : (b)
(b) Hepatitis B
(c) Heptatitis C
(d) Hepatitis E.

Q. 948. 'Cirrhosis' derived from the greek word meaning which one?

(a) Yellow **Answer** : (a)
(b) Granular liver
(c) Large liver
(d) Tumour.

Q. 949. Which one is an unique example of satellite virus disease in humans?

(a) Hepatitis A (b) Hepatitis B **Answer** : (d)
(c) Hepatitis C (d) Hepatitis D.

Q. 950. Which one is a DNA virus?

(a) Hepatitis B **Answer** : (a)
(b) Non A non B hepatitis (Hepatitis C)
(c) Hepatitis D
(d) Hepatitis E.

Q. 951. Hepatitis E infection could have devastating effects in which group?

(a) Children **Answer** : (b)
(b) Pregnant women
(c) Elderly
(d) Adolescents.

Q. 952. In which condition elevated CSF glutamine may be found?

(a) Schizophrenia **Answer** : (c)
(b) Cerebral stroke
(c) Hepatic encephalopathy
(d) Hysteria.

Q. 953. Ehrlich reagent is used to detect or measure which one?

(a) Urobilinogen
(b) Bilirubin
(c) Bile salt
(d) Urea.

Answer : (a)

Q. 954. Which one of these does not produce appreciable amounts of steroids?

(a) Testes
(b) Ovaries
(c) Placenta
(d) Liver.

Answer : (d)

Q. 955. The alpha subunit of which group is immunologically identical?

(a) TSH, HCG, FSH, LH
(b) TSH, GH, ACTH, PTH
(c) GH, ACTH, MSH, PTH
(d) TSH, FSH, GH, MSH.

Answer : (a)

Q. 956. Which protein have controlling effect on adenyl cyclase & phospholipase-C?

(a) P Protein
(b) G Protein
(c) I Protein
(d) Y Protein.

Answer : (b)

Q. 957. Which group of compounds produce the yellow pigment by Porter-silber method?

(a) Peptides
(b) Amino acids
(c) Steroids
(d) Prostaglandins.

Answer : (c)

Q. 958. Which is the classical procedure for quantitating meta & normetanephrines?

(a) Zimmerman reaction
(b) Porter-silber method
(c) Pisano method
(d) Flourescent method.

Answer : (c)

Q. 959. Vanillyl mandelic acid (VMA) is a catabolic compound of which group?

(a) Catecholamines
(b) Polypeptides
(c) Diamines
(d) Steroids.

Answer : (a)

Q. 960. Neurohypophysis is also known as :

(a) Hypothalamus
(b) Posterior pituitary
(c) Anterior pituitary
(d) Median lobe.

Answer : (b)

Q. 961. Prolactin inhibiting factor (PIF) is chemically which one of these?

(a) Serotonin **Answer** : (d)
(b) GABA
(c) Norepinephrine
(d) Dopamine.

Q. 962. On cleavage, beta lipotropin can produce which group?

(a) ACTH & MSH **Answer** : (b)
(b) MSH & endorphins
(c) MSH & melanin
(d) MSH & melatonin.

Q. 963. A patient has a serum T_4 of 12 mg/dl, the thyroxine hormone binding ratio (THBR) is 25% (normal = 30%). What is the FT_4I in this case?

(a) 6 **Answer** : (c)
(b) 8
(c) 10
(d) 16.

Q. 964. What could be the first and best laboratory test for primary thyroid disease?

(a) FT_4 **Answer** : (b)
(b) TSH
(c) THBR
(d) TBG.

Q. 965. Which is a more sensitive method to quantitate TSH?

(a) Spectro-photometric analysis **Answer** : (c)
(b) Immunometric analysis
(c) Radioimmuno assay
(d) Fluorimetric analysis.

Q. 966. Which of the following can be used as a postoperative marker in thyroid cancer?

(a) TSH **Answer** : (b)
(b) Thyroglobulin
(c) Reverse T_3
(d) FT_4.

Q. 967. Thyroperoxidase antibodies belong to which class of proteins?

(a) IgG **Answer** : (a)
(c) IgM
(c) IgA
(d) IgE.

Q. 968. Radioactive iodine uptake (RAIU) test is not very helpful in :

(a) Hyperthyroidism **Answer** : (b)
(b) Hypothyroidism
(c) Metastatic cancer of thyroid
(d) Malignancy of thyroid.

Q. 969. Which is not associated with hypothyroidism?

(a) Hypercholesterolemia
(b) Increased lipoprotein (a)
(c) Hypernatremia
(d) Hypoglycemia.

Answer : (c)

Q. 970. Defect of tertiary congenital hypothyroidism lies in :

(a) Thalamus
(b) Pituitary
(c) Hypothalamus
(d) Medulla oblongata.

Answer : (c)

Q. 971. Which is not true of TSH?

(a) TSH level may transiently rise in persons recovering from severe illness
(b) It may be temporarily high in newborns
(c) It is decressed by intake of amiodarone
(d) TSH secretion may be depressed by dopamine.

Answer : (c)

Q. 972. What percentage of thyrotoxicosis patients may develop the complication of 'thyroid storm'?

(a) 0·1%
(b) 0·5%
(c) 1%
(d) 5%.

Answer : (c)

Q. 973. Which may not be feature of thyroid storm?

(a) Congestive cardiac failure
(b) Confusion
(c) Insensitivity to pain
(d) Sharp rise in thyroxin level.

Answer : (d)

Q. 974. Infection of the mother during the 1st trimester by which virus could cause congenital heart disease in foetus?

(a) Hepatitis A
(b) Hepatitis B
(c) Rubella virus
(d) Varicella virus.

Answer : (c)

Q. 975. Which is most sensitive marker of acute myocardial infarction in first 2 hrs of onset of chest pain?

(a) AST
(b) CK-MB
(c) LDH
(d) Myoglobin.

Answer : (d)

Q. 976. Which one is extremely specific for acute myocardial injury?

(a) Troponin I (TnI)
(b) Tropinin C
(c) Heart fatty acid binding protein (H-FABP)
(d) Myosin light chains (MLL).

Answer : (a)

Q. 977. From among the choices given which is the most sensitive index of acute myocardial injury during an ischaemic attack?

(a) Glycogen phosphorylase BB (GPBB) **Answer** : (a)
(b) CK-MB
(c) Myoglobin
(d) TnT.

Q. 978. Which group of hormones is secreted by islet cells of pancreas?

(a) Insulin **Answer** : (d)
(b) Insulin & glucagon
(c) Insulin, glucagon & somatostatin
(d) Insulin, glucagon, gastrin & somatostatin.

Q. 979. What is the normal ratio of amylase clearance to creatinine clearance?

(a) Less than 3·1% **Answer** : (a)
(b) Between 3·5% to 5%
(c) Between 5 to 6·3%
(c) Always less than 1%.

Q. 980. Acute pancreatitis shows all the following abnormalities except?

(a) Increased triglyceride **Answer** : (d)
(b) Increased amylase
(c) Increased lipase
(d) Increased protein.

Q. 981. Which test can be used for direct determination of exocrine secretory capacity of pancreas?

(a) Fecal fat analysis **Answer** : (c)
(b) Amylase
(c) Secretin/CCK test
(d) Sweat chloride test.

Q. 982. Sweat chloride content greater than which amount is widely accepted as diagnostic of cystic fibrosis in children?

(a) 50 Meq/L **Answer** : (b)
(b) 60 Meq/L
(c) 70 Meq/L
(d) 80 Meq/L.

Q. 983. Which one tests only the absorptive ability of intestine?

(a) D-Xylose absorption test **Answer** : (a)
(b) Lactose tolerance test
(c) Serum carotenoids
(d) Serum amylase.

Q. 984. In which one these conditions gastrin may be greatly elevated?

(a) Gastric ulcer **Answer** : (b)
(b) Zollinger Ellison syndrome
(c) Perniceous anemia
(d) Achlorohydria.

Q. 985. Which is the major volume source of amniotic fluid in second half of pregnancy?

(a) Transudation across foetal skin **Answer :** (b)
(b) Fetal micturition
(c) Maternal secretion
(d) Secretions from placenta.

Q. 986. The tests for assessment of fetal lung maturity is based on which fact?

(a) Differentiation of alveolar epithelial cells to type I cells and their products **Answer :** (b)
(b) Differentiation of alveolar epithelial cells to type II cells and their products
(c) Production of AFP
(d) Production of meconium.

Q. 987. What is the total CSF volume?

(a) 50 ml (b) 100 ml **Answer :** (c)
(c) 150 ml (d) 200 ml.

Q. 988. On as average how much of CSF is formed per day?

(a) 200 ml **Answer :** (d)
(b) 300 ml
(c) 400 ml
(d) 500 ml.

Q. 989. What is 'Lamella bodies' of ammiotic fluid?

(a) Clumps of platelets of foetal origin **Answer :** (c)
(b) Clumps of platelets of maternal origin
(c) Packets of surfactant released by type 2 cells
(d) Clumps of fatty acid.

Q. 990. Which does not lead to a decrease in CSF glucose?

(a) Disorder in carrier mediated transport of glucose into CSF **Answer :** (d)
(b) Active metabolism of glucose by invading bacteria
(c) Increased metabolism in CNS due to neoplasms
(d) Viral infections of meninges.

Q. 991. Which is the standard method for diagnosis of cystic fibrosis?

(a) Identification of pancreatic insufficiency **Answer :** (b)
(b) Pilocarpine nitrate iontophoresis for sweat chloride measurements
(c) Presence of pneumonia
(d) Collapse of lungs.

Q. 992. Which is not a characteristic feature of synovial fluid?

(a) It is clear **Answer :** (d)
(b) It is colourless to pale yellow
(c) It is viscous in nature
(d) It can clot.

Q. 993. More than which amount of fluid in peritoneal cavity indicates diseases?

(a) 50 ml
(b) 100 ml
(c) 150 ml
(d) 200 ml.

Answer : (a)

Q. 994. Which is the most sensitive organ among most people to the toxic effects of alcohol?

(a) Brain
(b) Liver
(c) Kidney
(d) Intestine.

Answer : (b)

Q. 995. Among the long term and high level intake alcoholics, what percentage may develop cirrhosis?

(a) 10% (b) 20%
(c) 30% (d) 40%.

Answer : (b)

Q. 996. Which compound derived from methanol metabolism is said to be responsible for optic neuropathy?

(a) Acetaldehyde
(b) Formaldehyde
(c) Formic acid
(d) Acetic acid.

Answer : (c)

Q. 997. Consumption of which compound may lead to calcium oxalate crystal formation in kidney & renal tubular damage?

(a) Ethanol
(b) Methanol
(c) Isopropanol
(d) Ethylene glycol.

Answer : (d)

Q. 998. What is the primary end product of isopropanol ingestion so that it produces extended acute phase ethanol like symptom?

(a) Acetaldelyde
(b) Butyldehyde
(c) Acetone
(d) Propanal.

Answer : (c)

Q. 999. For each 60 mg/dl of increase in serum ethanol, osmolatlity of serum rise by how much?

(a) 10 m Osm/kg
(b) 20 m Osm/kg
(c) 30 m Osa/kg
(d) 5 m Osm/kg.

Answer : (a)

Q.1000. Which one is the reference method of serum ethanol determination?

(a) Spectrophotometry
(b) Gas chromatography
(c) Liquid chromatography
(d) Osmolality measurement.

Answer : (b)

Q.1001. Which statement is true for a weakly acidic toxin (pH = 4)?

(a) It is passively absorbed in the stomach (pH = 3·0)
(b) It is absorbed in intestine (pH = 8·5)
(c) It shall not be absorbed
(d) It requires a carrier or a weak base for absorption.

Answer : (a)

Q.1002. Which oncofetal antigen is helpful in correlating tumor burden in ovarian carcinoma?

(a) Carcinoembryonic antigen (CEA)
(b) Alpha fetoprotein
(c) CA 125
(d) CA 19-9.

Answer : (c)

Q.1003. Which is not true for prostatic specific antigen (PSA)?

(a) PSA is a serine protease
(b) PSA is organ specific & tumor specific
(c) PSA is complexed to alpha-I-antichymotrypsin in plasma
(d) PSA has about 90% sensitivity in stage D prostatic carcinoma.

Answer : (b)

Q.1004. Which one can be used as an index of placental function & foetal well-being?

(a) HCG
(b) Human placental lactogen
(c) Placental alkaline phosphatase
(d) CK—BB.

Answer : (b)

Q.1005. Which isoenzyme is raised in seminoma & teratomas?

(a) Alkaline phosphatase (heat labile)
(b) Acid phosphatase (tartrate labile)
(c) Lactate dehydrogenase—I
(d) Neuron specific enolase.

Answer : (c)

Q.1006. Which enzyme is used as a tumor marker?

(a) Lactate dehydrogenase
(b) Aldolase
(c) Carbonic anhydrase
(d) Lipase.

Answer : (a)

Q.1007. Measurement of DNA content (ploidy) is helpful in :

(a) Identifying type of cancer
(d) Detecting gene abnormality
(c) Discriminate between benign & malignant disease
(d) Immunophenotyping.

Answer : (c)

Q.1008. An isomer of which vitamin is used for treatment in acne?

(a) Vit. A
(b) Vit. C
(c) Vit. K
(d) Vit. E.

Answer : (a)

Q. 1009. Excess intake of which vitamin in pregnancy can lead to fetal malformation?

Answer : (a)

(a) Vit. A
(b) Folic acid
(c) Vitamin C
(d) Vitamin D.

Q. 1010. Elevation of urine or serum homocysteine is seen in deficiency of which vitamin?

Answer : (c)

(a) Riboflavin
(b) Thiamin
(c) Folic acid
(d) Biotin.

Q. 1011. Which vitamin shows enterohepatic circulation?

Answer : (b)

(a) Folic acid
(b) Vitamin B_{12}
(c) Riboflavin
(d) Thiamine.

Q. 1012. Which vitamin has an enediol group & can act as a reducing compound?

Answer : (a)

(a) Ascorbic acid
(b) Vitamin D
(c) Niacin
(d) Pyridoxine.

Q. 1013. Valproic acid therapy can cause increased urinary loss of which compound?

Answer : (b)

(a) Retinol
(b) Carnitine
(c) Tocopherol
(d) Pantothenic acid.

Q. 1014. What value of serum albumin indicates moderate degree of malnutrition?

Answer : (c)

(a) 35 g/L
(b) 28 g/L
(c) 23 g/L
(d) 15 g/L.

Q. 1015. In which condition transferrin level may be decreased?

Answer : (d)

(a) Pregnancy
(b) Acute hepatitis
(c) Birth control pills
(d) End stage liver disease.

Q. 1016. Which protein migrates ahead of albumin in electrophoresis?

Answer : (a)

(a) Transthyretin
(b) Transferrin
(c) Fibrinonectin
(d) Haptoglobin.

Q. 1017. Which protein is a better marker of visceral protein synthesis in patients receiving nutritional support?

(a) Albumin
(b) Transferrin
(c) Transthyretin
(d) Fibrinogen.

Answer : (c)

Q. 1018. When a patient is put on total parenteral nutrition (TPN) on long term basis, which trace element should be monitored on a weekly basis?

(a) Copper
(b) Selenium
(c) Molybdenum
(d) Manganese.

Answer : (a)

Q. 1019. Which serum analyte increases with aging?

(a) Albumin
(b) Lactate dehydrogenase
(c) Haptoglobin
(d) Growth hormone.

Answer : (b)

Q. 1020. Which serum analyte decreases with aging?

(a) Glucose (fasting)
(b) Triglyceride
(c) Amylase
(d) Bilirubin.

Answer : (d)

Q. 1021. Which is not true in elderly?

(a) Creatine clearance is decreased
(b) There is decrease sensation of thirst
(c) Dehydration is a common & under appreciated condition
(d) BUN decreases.

Answer : (d)

Q. 1022. Which analyte remains unchanged in the elderly?

(a) Fasting glucose
(b) BUN
(c) Haptoglobin
(d) Transferrin.

Answer : (c)

Q. 1023. What is the reference range of triglyceride in infants (less than 1yr of age)?

(a) 20—120 mg/dl
(b) 50—190 mg/dl
(c) 70—190 mg/dl
(d) 25—270 mg/dl.

Answer : (a)

Q. 1024. Which is the major antibody synthesized by human foetus?

(a) IgG
(b) IgM
(c) IgA
(d) IgE.

Answer : (b)

Q. 1025. A physiological hypogammaglobulinemia may occur in infants during which period of life?

(a) 1—2 month
(b) 2—4 month
(c) 4—9 month
(d) 9—12 month.

Answer : (c)

Q. 1026. By which method products of PCR analysis can be separated?

(a) Capillary electrophoresis
(b) TLC
(c) Ion exchange chromatography
(d) Cellulose acetate electrophoresis.

Answer : (a)

Q. 1027. Which is the most common radionucleotide with gamma emission used in laboratory for labelled assay?

(a) ^{3}H
(b) ^{125}I
(c) ^{129}I
(d) ^{131}I.

Answer : (b)

Q. 1028. Following an acute myocardial infarction cTnI rises to a peak in :

(a) 1—2 hrs
(b) 3—6 hrs
(c) 8—10 hrs
(d) 14—20 hrs.

Answer : (d)

Q. 1029. What is true for lactate dehydrogenase—6 (LD—6)?

(a) There is no LD—6
(b) LD—6 is skeletal muscle specific
(c) LD—6 is neurone specific
(d) LD—6 may reflect liver injury secondary to circulatory insufficiency.

Answer : (d)

Q. 1030. Hydroxy butyrate dehydrogenase (HBD) activity increases in conditions in which :

(a) There is LD—1 & LD—2 rise
(b) There is LD—3 & LD—4 rise
(c) There is LD—5 rise
(d) It has got no association with rise of LD.

Answer : (a)

Q. 1031. Persons with which blood group may show presence of intestine alkaline phosphatase isoenzyme in serum?

(a) A
(b) B
(c) AB
(d) None of the above.

Answer : (b)

Q. 1032. The Nagao isoenzyme of alkaline phosphatase may be a variant of which one?

(a) Liver isoenzyme of ALP
(b) Bone isoenzyme of ALP
(c) Intestinal isoenzyme of ALP
(d) Regan isoenzyme of ALP.

Answer : (d)

Q. 1033. Which is the more accurate method of plasma glucose estimation?

(a) Glucose oxidase
(b) Hexokinase
(c) Somogyi Nelson method
(d) Folin—wu method.

Answer : (b)

Q. 1034. Which enzyme is used in the coupling reaction of hexokinase method of plasma glucose estimation?

(a) Glucose–6–phosphate dehydrogenase
(b) Glucose–6–phosphate
(c) Hexokinase
(d) Phospho-glucomutase.

Answer : (a)

Q. 1035. How much glucose is given orally to children weighing less than 40 Kg in oral glucose tolerance test?

(a) 1 gm/kg body weight
(b) 1·25 gm/kg body weight
(c) 1·5 gm/kg body weight
(d) 1·75 gm/kg body weight.

Answer : (d)

Q. 1036. Serial measurement of blood ammonia could be helpful in which condition?

(a) Acid base disorders
(b) Acute tubular necrosis
(c) Liver failure
(d) Congestive cardiac failure.

Answer : (c)

Q. 1037. Which the most common cause of hypermagnesemia?

(a) Digitalis therapy
(b) Renal failure
(c) Chronic laxative use
(d) Hyperparathyroidism with hypercalcemia.

Answer : (b)

Q. 1038. Which is the principal route of copper excretion from the Human body?

(a) Urine
(b) Sweat
(c) Bile
(d) Faeces.

Answer : (c)

Q. 1039. Which element competes with copper for absorption from intestine?

(a) Sulphur
(b) Zinc
(c) Molybdenum
(d) Magnesium.

Answer : (b)

Q. 1040. The X-linked recessive disorder, Menke's syndrome, is associated with deficiency of which element?

(a) Zinc
(b) Iron
(c) Selenium
(d) Copper.

Answer : (d)

Q. 1041. In Kashin Beck disease, the necrosis of cartilage is attributed to deficiency of which element?

(a) Molybdenum **Answer** : (b)
(b) Selenium
(c) Zinc
(d) Copper.

Q. 1042. In blood gas analyzers measurement of which one is amperometric?

(a) pO_2 **Answer** : (a)
(b) pCO_2
(c) pH
(d) HCO_3.

Q. 1043. The electrodes of which parameter is called Clarke's electrodes?

(a) pO_2 **Answer** : (a)
(b) pCO_2
(c) pH
(d) $HCO^-{}_3$.

Q. 1044. Which is not associated with Verner—Morrison syndrome?

(a) Watery diarrhoea **Answer** : (c)
(b) Dehydration
(c) Hyperchlorhydria
(d) Increased levels of VIP.

Q. 1045. Which is not true of Multiple Endocrine neoplasia type II, also known as Sipple's syndrome?

(a) Autosomal recessive inheritence **Answer** : (a)
(b) Medullary carcinoma of thyroid
(c) Pheochromocytoma
(d) Parathyroid adenoma or hyperplasia.

Q. 1046. The alpha subunit of hCG is not similar to a subunit of which one?

(a) FSH **Answer** : (d)
(b) LH
(c) TSH
(d) GH.

Q. 1047. In critically ill patients with multi-organ failure, which is a preferred marker of myocardial injury?

(a) CK—MB **Answer** : (b)
(b) Troponin I
(c) Lactate dehydrogenase—I
(d) Cardiac myosin heavy chains.

Q. 1048. Production of which one of these compounds is regulated by blood oxygen levels?

(a) Renin **Answer** : (c)
(b) Prostaglandins
(c) Erythropoetin
(d) Angiotensinogen.

Q. 1049. GFR should reduce to what amount to define this condition as acute renal failure?

(a) 40 ml/min **Answer** : (d)
(b) 30 ml/min
(c) 20 ml/min
(d) 10 ml/min.

Q. 1050. Which type of hyperlipemia is most prevalent in patients on chronic dialysis?

(a) Type I **Answer** : (c)
(b) Type II
(c) Type IV
(d) Type V.

Q. 1051. In normal adults after an oral load of 25g of D-xylose during D-xylose test, what is the least amount of xylose excreted in urine in the 5hr period?

(a) 1 gm (b) 2 gms **Answer** : (c)
(c) 4 gms (d) 6 gms.

Q. 1052. Which is the first disease to be recognised by laboratory procedures of amniotic fluid?

(a) Neural tube defects **Answer** : (b)
(b) Erythroblastosis foetalis
(c) Thalassemia major
(d) Cirrhosis of liver.

Q. 1053. Intake of which toxic substance typically presents with renal tubular dysfunction as an early finding?

(a) Cyanide **Answer** : (c)
(b) Arsenic
(c) Cadmium
(d) Lead.

Q. 1054. Which one is most widely investigated and most frequently used tumor marker in cancer management?

(a) Prostate specific antigen (PSA) **Answer** : (b)
(b) Carcino embryonic antigen (CEA)
(c) Alpha fetoprotein
(d) CA125.

Q. 1055. Which is not an oncofetal antigen?

(a) Carcino embryonic antigen **Answer** : (d)
(b) Alpha fetoprotein
(c) Prostate specific antigen
(d) Neuron specific enolase.

Q. 1056. Which one of these is complexed to alpha–I–antichymotrypsin in blood?

(a) Prostate specific antigen **Answer** : (a)
(b) Tissue polypeptide antigen
(c) Human chorionic gonadotropin
(d) Placental alkaline phosphatase.

Q. 1057. Which one of these is typically elevated in about 70% cases of small carcinoma of lungs?

Answer : (b)

(a) Acid phosphatase
(b) Neuron specific enolase
(c) CA 549
(d) CK–BB.

Q. 1058. Which one is also known as insulin growth factor—I?

Answer : (a)

(a) Somatomedin—C
(b) Sometomedin—D
(c) Somatostatin—A
(d) Somatostatin—A.

Q. 1059. Which one can be used as a reliable index in monitoring short term nutritional support in infants with protein malnutrition?

Answer : (c)

(a) C–reactive protein
(b) Albumin
(c) Fibronectin
(d) Retinol binding protein.

Q. 1060. Which are the two most common trace elements added to total parenteral nutrition (TPN) solution?

Answer : (b)

(a) Magnesium & zinc
(b) Zinc & copper
(c) Copper & calcium
(d) Calcium & iodine.

REFERENCE READING

1. **Harper's Biochemistry**—25th and 26th Edition
2. **Lecture Notes on Clinical Biochemistry**—6th Edition (Blackwell Science Publication)
3. **Tietz Text Book of Clinical Chemistry**—3rd Edition
4. **Bishop's Clinical Chemistry**—Principles, Procedures, Correction, 4th Edition (Lippincort, William & Wilkins Publication)